Marine Pollution

Marine Pollution

THIRD EDITION

R. B. Clark

Emeritus Professor of Zoology
University of Newcastle upon Tyne

CLARENDON PRESS · OXFORD
1992

Oxford University Press, Walton Street, Oxford OX2 6DP
Oxford New York Toronto
Delhi Bombay Calcutta Madras Karachi
Petaling Jaya Singapore Hong Kong Tokyo
Nairobi Dar es Salaam Cape Town
Melbourne Auckland
and associated companies in
Berlin Ibadan

Oxford is a trade mark of Oxford University Press

Published in the United States
by Oxford University Press, New York

First published 1986
Second edition 1989
Third edition 1992

A catalogue record of this book is available from the British Library

Library of Congress Cataloging in Publication Data
Clark, R. B. (Robert Bernard), 1923–
Marine pollution/R. B. Clark.—3rd ed.
Includes bibliographical references and index.
1. Marine pollution. I. Title.
GC1085.C5 1992   363.73'94—dc20   91–41042
ISBN 0-19-854686-6
ISBN 0-19-854685-8

Typeset by Cotswold Typesetting Ltd, Cheltenham
Printed in Great Britain by Bookcraft (Bath) Ltd,
Midsomer Norton, Avon

PREFACE TO THE THIRD EDITION

This textbook stems from an introductory course of lectures on marine pollution that I gave for some years to marine biologists in the University of Newcastle upon Tyne. Designing a lecture course (and writing a textbook) on this subject has become increasingly difficult in recent years.

It is now six years since the first edition of this book was published and almost ten years since it first began to take shape. The most notable change during that decade has been in public opinion, not in science. In the early 1980s, there were voices proclaiming that the sea was being poisoned by pollution; indeed, some claimed that the marine environment had already been irreparably damaged by our waste discharges. To a considerable extent, these pessimistic voices were disregarded and dumping of wastes into the sea continued on a large scale in most parts of the world. Now, environmental expectations are high and continue to rise, and waste disposal practices have had to be adjusted to take account of this change of public attitude.

The objective of university teaching is partly to give students access to sources of current information but also, importantly, to encourage them to think independently and form scientific judgements which they can justify and support with tested evidence.

The difficulty with a subject such as environmental pollution is that students almost invariably already hold strong views about it. These are not views that they have developed themselves by a rational examination of the facts, but usually simply reflect current conventional wisdom. The conventional view has changed radically in the last ten years even though the scientific basis for it has not. The current public assessment may be right or wrong, but it and the grounds on which it was based must not be above dispassionate examination. However, for many environmental enthusiasts, dispassionate examination of the facts is either irrelevant or, worse, indicates an 'anti-environmental' attitude. In these circumstances, the teacher, whether he subscribes to the conventional view or not, is in danger of being accused of trying to indoctrinate the students, and this is not a happy situation when the objective of the instruction is to encourage rational and independent thought.

The scientific appraisal of the health of the seas and the principal threats to the marine environment made a decade ago have proved surprisingly reliable and, except in matters of detail, have not changed appreciably in that time. Estimates of the quantity of wastes entering the sea and the environmental damage they cause have steadily reduced in most parts of the world. There have also been some changes in priorities; discharges of untreated sewage, for example, are a major health risk in many tropical countries and are now regarded much more seriously than oil, pesticides, or heavy metals in the catalogue of damaging additions to the sea.

Chapters 2 to 8 give a purely factual account of the quantities of wastes of different types entering the sea, where they come from, and how they reach the sea; what happens to them in the sea; their toxicity as revealed in laboratory studies, and, more importantly, the observed effects they have in the marine environment; and the threat they may pose to human health. The case-histories that have been used to illustrate pollution effects are generally the same as those used in previous editions, but throughout these chapters data have been updated to reflect the most recent accepted values. Some waste disposal practices have now been abandoned or are being phased out, but these are still discussed in the appropriate chapters because we still have to deal with the legacy of damage that they have caused. Some events, such as the nuclear explosion at Chernobyl, were not treated in previous editions but have now been added.

Chapter 9 considers the special problems of a number of sea areas. These sections have been updated where necessary, but since the treatment is necessarily rather general, these chapters have required much less modification than the others.

Chapters 1 and 10 may be controversial, not for what they say, but because I have been at pains to avoid offering judgements. Chapter 1 explains the treatment given the subject in the following chapters, and emphasizes the need for an objective, step-by-step approach to examining the consequences of waste disposal in the sea. It then suggests that neither pollution nor its abatement are as simple as might appear at first sight. Chapter 10 enlarges on this and explores the difficulties that have arisen in the scientific study of marine pollution. Readers are challenged to form their own judgements and to justify their conclusions by reference to factual evidence.

Some reviewers of previous editions have regretted the lack of detailed references to the authority for statements in the book, but these would be inappropriate in a textbook. The section of Acknowledgements gives details of the sources of all the text figures and tables, and these will almost always guide the interested reader to the relevant literature.

Newcastle upon Tyne   R. B. C.
January 1992

ACKNOWLEDGEMENTS

Professor J. S. Gray (Oslo) and Professor S. A. Gerlach (Kiel) made critical and helpful comments on the previous edition of this book, for which the author is most grateful. Most of their suggestions have been incorporated in this edition, but they cannot be held responsible for my treatment of them. Mr David Moulder (Plymouth Marine Laboratory, Plymouth) helped compile the section on Further Reading.

The author also gratefully acknowledges the following sources of figures and data in this book. **Fig. 1.1** Royal Commission on Environmental Pollution, *Tenth Report*, HMSO, London (1984); **Fig. 2.4** Royal Commission on Environmental Pollution, *Third Report*, HMSO, London (1972); **Fig. 2.5** Royal Commission on Environmental Pollution, *Third Report*, HMSO, London (1972); **Fig. 2.7** DSIR Wat. Pollut. Res. Pap., No. 11, *Effects of polluting discharges on the Thames estuary*, HMSO, London (1964); **Fig. 2.8** Royal Commission on Environmental Pollution, *Third Report*, HMSO, London (1972); **Fig. 2.9** (a), (d) House of Lords Select Committee on the European Communities, Session 1985-6, *17th Rept*, (HL 219), HMSO, London (1986); (b), (c) Mackay, D. W. *et al.*, *Mar. Pollut. Bull.*, **3**, 7-10 (1973); **Fig. 2.10** House of Lords Select Committee on the European Communities, Session 1985-6, *17th Rept*, (HL 219), HMSO, London (1986); **Fig. 2.11** Shelton, R. G. J., *Mar. Pollut. Bull.*, **2**, 24-7 (1971); **Fig. 2.12** (a), (c), (d) from: Pearce, J. B. *et al.*, *Mar. Pollut. Bull.*, **5**, 132-3 (1974); (b) from: Pearce, J. B. and Young, J. S., *Mar. Pollut. Bull.*, **6**, 101-5 (1975); **Fig. 2.13** Caspers, H., *Proc. Wat. Tech., Toronto*, **12**, 461-79 (1980) and ap Rheinallt, T., in Salomons, W. *et al.*, *Pollution of the North Sea: an assessment*, Springer, Berlin (1988), 489-509; **Fig.**

2.14 Steimle, F. W. and Sindermann, C. J., *Mar. Fish. Rev.*, **40**, 17-26 (1978); **Fig. 2.15** Gerlach, S. A., *Ber. Inst. Meeresk.*, **130**, 1-87 (1984); **Fig. 2.16** Justić, D., *Mar. Pollut. Bull.*, **18**, 281-4 (1987); **Fig. 3.1** *Shell Briefing Service*, No. 1, Shell Petroleum International, London (1989); **Fig. 3.2** Wardley-Smith, J., *The control of oil pollution on the sea and inland water*, Graham and Trotman, London (1976); **Fig. 3.3** Sanders, P. F. and Tibbetts, P. J. C., *Phil. Trans. R. Soc. Lond.*, **B316**, 568 (1987); **Fig. 3.5** Whittle, K. J. *et al.*, *Phil. Trans. R. Soc. Lond.*, **B297**, 193-218 (1982); **Fig. 3.6** Wardley-Smith, J., *The control of oil pollution on the sea and inland waters*, Graham and Trotman, London (1976); **Fig. 3.7** Wardley-Smith, J., *The control of oil pollution on the sea and inland waters*, Graham and Trotman, London (1976); **Fig. 3.8** Wardley-Smith, J., *The control of oil pollution on the sea and inland water*, Graham and Trotman, London (1976); **Fig. 3.9** North, W. J. *et al.*, *Symp. Pollut. mar. Micro-org. Prod. pétrol.*, 335-54, CIESMM, Monaco (1964) and Nelson-Smith, A., *Oil pollution and marine ecology*, Elek, London (1972); **Fig. 3.10** Bourne, W. R. P., in *Marine pollution* (ed. R. Johnston), 403-502, Academic Press, London (1976); **Fig. 3.11** Dicks, B. and Hartley, J. P., *Phil. Trans. R. Soc. Lond.*, **B297**, 285-307 (1982); **Fig. 3.12** Dicks, B. and Hartley, J. P., *Phil. Trans. R. Soc. Lond.*, **B297**, 285-307 (1982); **Fig. 3.13** Dicks, B. and Hartley, J. P., *Phil. Trans. R. Soc. Lond.*, **B297**, 285-307 (1982); **Fig. 3.14** Kingston, P. F., *Phil. Trans. R. Soc. Lond.*, **B316**, 556 (1987); **Fig. 3.15** Kingston, P. F., *Phil. Trans. R. Soc. Lond.*, **B316**, 549 (1987); **Fig. 4.1** Brown, B. E., *Effects of heavy metals on the physiology of selected invertebrates*, unpublished thesis, University of London (1972); **Fig. 4.2** Bryan, G. W., in *Marine pollution* (ed. R. Johnston), 185-302, Academic Press, London

(1976); **Fig. 4.3** Coombs, T. L. and George, S. G., in *Physiology and behaviour of marine organisms* (ed. D. S. McClusky and A. J. Berry), 179–87, Pergamon, Oxford (1977); **Fig. 4.4** Bryan, G. W., in *Marine pollution* (ed. R. Johnston), 185–302, Academic Press, London (1976); **Fig. 4.5** Sunila, I., *Ann. Zool. Fenn.*, **18**, 213–23 (1981); **Fig. 4.6** (a) from: Anderson, D. S., *Embryology and phylogeny in annelids and arthropods*, Pergamon, Oxford (1973); (b) from: Reish, D. J. *et al.*, *Mar. Pollut. Bull.*, **5**, 125–6 (1974); **Fig. 5.1** Cunningham, F. A. and Tripp, M. R., *Mar. Biol.*, **31**, 311–19 (1975); **Fig. 5.2** Ünlü, M. Y. *et al.*, in *Marine pollution and sea life* (ed. M. Ruivo), 292–5, Fishing News (Books) Ltd., London (1972); **Fig. 5.3** Häkkinen, I. and Häsänen, E., *Ann. Zool. Fenn.*, **17**, 131–9 (1980); **Fig. 5.4** Korringa, P., *Helgol. wiss. Meeresunters.*, **17**, 126–40 (1968); **Fig. 5.5** Bryan, G. W. and Gibbs, P. E., *Occ. Pap. mar. biol. Ass. UK*, (2) (1983); **Fig. 5.6** Bryan, G. W., in *Marine pollution* (ed. R. Johnston), 185–302, Academic Press, London (1976); **Fig. 5.7** Murozumi, M. *et al.*, *Geochim. Cosmochim. Acta*, **33**, 1247–94 (1969); **Fig. 5.8** Lee, J. A. and Tallis, J. H., *Nature, Lond.*, **245**, 216–18 (1973); **Fig. 5.9** Gerlach, S. A., *Marine pollution: diagnosis and therapy*, Springer, Heidelberg (1981); **Fig. 5.10** Newell, R. C. *et al.*, *Mar. Pollut. Bull.*, **22**, 112–18 (1991); **Fig. 6.2** Smokler, P. E. *et al.*, *Mar. Pollut. Bull.*, **10**, 331–4 (1979); **Fig. 6.3** Halcrow, W. *et al.*, *Mar. Pollut. Bull.*, **5**, 134–6 (1974); **Fig. 6.4** Goerke, H. *et al.*, *Mar. Pollut. Bull.*, **10**, 127–33 (1979); **Fig. 6.5** Ernst, W. and Georke, H., *Mar. Biol.*, **24**, 287–304 (1974); **Fig. 6.6** Mellanby, K., *Pesticides and pollution*, Collins, London (1967); **Fig. 6.7** Patin, S. L., *Pollution and the biological resources of the oceans*, Butterworth, London (1982); **Fig. 6.8** Ernst, W., *Helgol. Meeresunters.*, **33**, 301–12 (1980); **Fig. 6.9** Bonner, W. N. and Hickling, G., *Trans. nat. Hist. Soc. Northumbria*, **42**, 65–84 (1974); **Fig. 7.2** Hodge, V. F. *et al.*, *Proc. Symp. Radioactive Contamination of the Marine Environment*, 263–76, IAEA, Vienna (1973); **Fig. 7.3** Kautsky, W., *Umschau*, **77**, 672–3 (1973); **Fig. 7.4** Radioactive Waste Management Committee, *Fourth Annual Report*, HMSO, London (1983), with additional data from *MAFF Aquatic Environment Monitoring Report*, **16–23** (1985–91); **Fig. 7.5** Hunt, G. J., *Aquatic Environment Monitoring Report*, **21**, Ministry of Agriculture, Fisheries and Food, Lowestoft (1989); **Fig. 7.6** Fowler, S. W. and Gaury, J. C., *Nature, Lond.*, **266**, 827 (1977) and Pentreath, R. J., *Nuclear power: man and the environment*, Taylor and Francis, London (1980); **Fig. 7.7** Woodhead, D. S., *Helgol. Meeresunters.*, **33**, 122–37 (1980); **Fig. 7.8** data from Webb, G. A. M. *et al.*, *National Radiological Protection Board, Rept* NRPB-M286, Didcot, UK (1991); **Fig. 8.1** Rosenberg, R., *Mar. Pollut. Bull.*, **8**, 102–4 (1977); **Fig. 8.2** Anon., *Mar. Pollut. Bull.*, **1**, 115 (1970); **Fig. 8.3** (a) from: Portmann, J. E., *J. mar. biol. Ass. UK*, **50**, 577–91 (1970); (b), (c) from: Howell, B. R. and Shelton, R. G. J., *J. mar. biol. Ass. UK*, **50**, 593–607 (1970); **Fig. 8.4** Dixon, T. R. and Dixon, T. J., *Mar. Pollut. Bull.*, **10**, 352–7 (1979); **Fig. 8.5** Thorhaug, A., *Mar. Pollut. Bull.*, **9**, 191–7 (1978); **Fig. 8.6** Langford, T. E., *Electricity generation and the ecology of natural waters*, University of Liverpool Press, Liverpool (1983); **Fig. 9.1** Korringa, P., *Helgol. wiss. Meeresunters.*, **17**, 126–40 (1968); **Fig. 9.2** (a) Royal Commission on Environmental Pollution, *Tenth Report*, HMSO, London (1984); (b)–(e) based on data in *Quality Status of the North Sea*, Dept of Environment (1987); **Fig. 9.3** Heimer, R., *Ambio*, **6**, 313–16 (1977); **Fig. 9.4** Le Lourd, P., *Ambio*, **6**, 317–19 (1977); **Fig. 9.6** Fonselius, S., *Mar. Pollut. Bull.*, **12**, 187–94 (1981); **Fig. 9.7** Andersin, A. B., *Kieler Meeresforsch.*, Suppl. **4**, 23–52 (1978); **Fig. 9.8** Thorhaug, A., *Ambio*, **10**, 295–8 (1981); **Fig. 9.9** Beelhuis, J. V., *Ambio*, **10**, 325–31 (1981); **Fig. 9.10** Adapted from: Zenkevich, L. A., *Treatise on marine ecology and paleoecology*, **1**, 891–916, Geol. Soc. Am., New York (1957); **Fig. 10.1** Mann, K. H., *Helgol. wiss. Meeresunters.*, **30**, 445–67 (1977); **Fig. 10.2** Lewis, J. R., *Phil. Trans. R. Soc. Lond.*, **B297**, 257–67 (1982); **Fig. 10.3** Lewis, J. R., in *Marine pollution and sea life* (ed. M. Ruivo), 401–4, Fishing News (Books) Ltd., London (1972); **Fig. 10.4** Bowman, R. S. and Lewis, J. R., *J. mar. biol. Ass. UK*, **57**, 793–815 (1977); **Fig. 10.5** Buchanan, J. B. *et al.*, *J. mar. biol. Ass. UK*, **58**, 191–209 (1978); **Fig. 10.6** Glover, R. S. *et al.*, in *Marine pollution and sea life* (ed. M. Ruivo), 439–45, Fishing News (Books) Ltd., London (1972); **Fig. 10.7** Jones, R., *Phil. Trans. R. Soc. Lond.*, **B297**, 353–68 (1982); **Fig. 10.8** Jones, R., *Phil. Trans. R. Soc. Lond.*, **B297**, 353–68 (1982); **Fig. 10.9** Battaglia, B., *Helgol. Meeresunters.*, **33**, 587–95 (1980); **Fig. 10.10** Mann, K. H., *Mar. Biol.*, **14**, 119–209 (1972). **Table 3.1** Steering Committee for the Petroleum in the Marine Environment Update, *Oil in the sea: inputs, fates and effects*, National Academy Press, Washington, DC (1985) and *Mar. Pollut. Bull.*, **22**, 262 (1991); **Table 3.2** *Our industry petroleum*, BP Ltd., London (1977);

Table 3.3 Dicks, B. and Hartley, J. P., *Phil. Trans. R. Soc. Lond.*, **B297**, 285–307 (1982); **Table 3.4** Royal Commission on Environmental Pollution, *Eighth Report*, HMSO, London (1981); **Fig. 3.5** Royal Commission on Environmental Pollution, *Eighth Report*, HMSO, London (1981); **Table 4.1** Bryan, G. W. *et al.*, *Mar. biol. Assoc. U.K., Occ. Publ.*, (4) (1985); **Table 4.2** White, H. H. and Champ, M. A., in *Symp. Hazardous and Industrial Solid Waste Testing* (ed. R. A. Conway and W. P. Gulledge), 299–312, Amer. Soc. Testing and Materials, Philadelphia (1983); **Table 4.3** Gray, J. S. and Ventilla, R. J., *Ambio*, **2**, 118–211 (1973); **Table 4.4** Bryan, G. W., in *Marine Pollution* (ed. R. Johnston), 185–230, Academic Press, London (1976); **Table 5.1** Scientific and Technical Working Group, *Quality status of the North Sea: summary*, Department of the Environment, London (1987); **Table 5.2** Pacyna, J. M., in *Toxic metals in the atmosphere* (ed. J. O. Nriagu and C. I. Davidson), 33–52, Wiley Interscience, London (1986); **Table 5.3** Buat-Ménard, P., in *The role of air–sea interchange in geochemical cycling* (ed. P. Buat-Ménard), 477–496, Reidel, Dordrecht (1986); **Table 5.4** Keckes, S. and Miettinen, J. K., in *Marine pollution and sea life* (ed. M. Ruivo), 276–89, Fishing News (Books) Ltd., London (1972) and Craig, P. J., *Organometallic compounds in the environment principles and reactions*, Longman, London (1986); **Table 5.5** Keckes, S. and Miettinen, J. K., in *Marine pollution and sea life* (ed. M. Ruivo), 276–89, Fishing News (Books) Ltd., London (1972); **Table 5.6** Boney, A. D. *et al.*, *Biochem. Pharmacol.*, **2**, 37–49 (1959); **Table 5.7** Häkkinen, I. and Häsänen, E., *Ann. Zool. Fenn.*, **17**, 131–9 (1980); **Table 6.1** Royal Commission on Environmental Pollution, *Tenth Report*, HMSO, London (1984); **Table 6.2** House of Lords Select Committee on European Communities, Session 1987–88, *17th Rept*, (HL219), HMSO, London (1988); **Table 6.3** Atlas, E. L. and Giam, C. S., *Science*, **211**, 163–5 (1981); **Table 6.4** Riseborough, R. W. *et al.*, *Mar. Pollut. Bull.*, **7**, 225–8 (1976); **Table 6.5** Butler, P. A., in *Marine pollution and sea life* (ed. M. Ruivo), 262–6, Fishing News (Books) Ltd., London (1972); **Table 6.6** Smokler, P. E. *et al.*, *Mar. Pollut. Bull.*, **10**, 331–4 (1979); **Table 7.1** International Atomic Energy Authority, *Technical Reports Series*, **172**, IAEA, Vienna (1976) units adjusted; **Table 7.2** Woodhead, D. S., *Proc. Symp. Radioactive Contamination in the Marine Environment*, 455–525, IAEA, Vienna (1976) units adjusted; **Table 7.3** *Report of the Independent Review of Disposal of Radioactive Waste in the Northeast Atlantic*, HMSO, London (1984) units adjusted; **Table 7.4** International Atomic Energy Authority, *Technical Report Series*, **172**, IAEA, Vienna (1976) units adjusted; **Table 7.5** Woodhead, D. S., in *Marine Ecology* (ed. O. Kinne) **5**, 1112–287, Wiley, London (1984) units adjusted; **Table 7.6** Camplin, W. C. and Aarkrog, A., *Data Rep., MAFF Direct. Fish. Res., Lowestoft*, No. 20; (1989). **Table 8.1** Oslo Commission *Eleventh Annual Report* (1986); **Table 8.2** Dixon, T. R. and Dixon, T. J., *Mar. Pollut. Bull.*, **10**, 352–7 (1979); **Table 9.1** Scientific and Technical Working Group, *Quality status of the North Sea: summary*, Department of the Environment, London (1987); **Table 9.2** Kullenberg, G., in *The role of the oceans as a waste disposal option* (ed. G. Kullenberg), 388, Riedel, Dordrecht (1986); **Table 9.3** Brigmann, L., *Mar. Pollut. Bull.*, **12**, 214–18 (1981); **Table 9.4** Thorhaug, A., *Ambio*, **10**, 295–8 (1981).

CONTENTS

1

WHAT IS POLLUTION?

SOME QUESTIONS

Everyone knows what pollution is and that it is a 'bad thing', but for a scientific examination of marine pollution, or any other sort of pollution, value judgements of this kind have to be quantified. In what way is it bad? How bad? Bad for whom?

To answer these questions, we must consider:

(1) what kind of materials are discharged into seas or estuaries, or otherwise get there as a result of human activities;

(2) what effect these additions to the sea have on the marine or estuarine environment and the plants and animals living there;

(3) what implications these effects have for human health, food resources, commercial interests, amenities, wildlife conservation, or ecosystems in general;

(4) what is being done, can be done, or should be done to reduce or remove the damaging or undesirable effects of these additions to the marine environment;

(5) what would be the consequences of not releasing these materials to the sea and would such consequences be better or worse than the existing situation.

A quantitative, step-by-step approach to this subject is clearly required, and the rest of this book is concerned with answering these questions in quantitative terms and with assessing the importance of various kinds of pollution impact. Before embarking on that, it is necessary to look at what types of materials enter the sea as a result of human activities.

CATEGORIES OF ADDITIONS

Degradable wastes

By far the greatest volume of discharge into coastal waters and estuaries is composed of organic material which is subject to bacterial attack. Essentially, this is an oxidative process and ultimately breaks down organic compounds to stable inorganic compounds such as CO_2 (carbon dioxide), H_2O (water), and NH_3 (ammonia).

Wastes included under this heading are:

(1) a large part of urban sewage;

(2) agricultural wastes (now in substantial quantities where factory farming is practised);

(3) food processing wastes from slaughter houses and freezer plants, pulp from sugar beet factories, and so on;

(4) brewing and distillery wastes;

(5) paper pulp mill wastes which include much wood fibre;

(6) chemical industry wastes, including a great variety of large molecules which are relatively unstable and readily broken down;

(7) oil spillages.

In principle, such degradable wastes are no different from plant and animal remains which are subject to bacterial decay. Since bacteria are an important base of many food chains in the sea, the addition of organic matter represents an enrichment of the eco-system, comparable to adding stable manure as a fertilizer in the garden.

If the rate of input exceeds the rate of bacterial degradation, organic materials **accumulate**. The rate of bacterial action depends on temperature, oxygen availability, and other factors; if these become limiting, the rate of bacterial action falls and the capacity of the waters to receive organic wastes without accumulation is much reduced. If the input of wastes is large, there is intense bacterial activity until the oxidative processes of degradation outrun the supply of oxygen dissolved in the water, leading to **deoxygenation**. In these circumstances, further degradation depends on the activity of **anaerobic** bacteria, which is slow and yields end-products such as hydrogen sulphide and methane.

Accumulation of organic materials and deoxygenation of the water both have a strong impact on the flora and fauna, and at very low oxygen levels most plants and animals are excluded. Thus, if the input of organic wastes is within the capacity of the receiving waters—which is related to temperature, oxygen avail-ability, water currents, and so on—it will result in enrichment, of benefit chiefly to plants in the first instance. If the capacity of the receiving waters is exceeded, the accumu-lation of organic material and the develop-ment of anoxic conditions results in impover-ishment of the fauna and flora.

Fertilizers

Agricultural fertilizers may have a similar effect to organic wastes. Nitrates and phos-phates are leached from arable land and carried by rivers to the sea, where the fertilizers enhance phytoplankton produc-tion, sometimes to the extent that the accumu-lation of dead plant-remains on the seabed produces anoxic conditions.

Dissipating wastes

A number of industrial discharges into the sea and estuaries rapidly lose their damaging properties after they enter the water. Any effects they may have are therefore confined to the area immediately around the point of discharge, though the extent of that area depends on the rate of discharge, water currents, and so on.

1. *Heat.* This comes principally from the cooling water from coastal power stations and factories, although some other indus-trial effluents may be heated. Commonly the discharge is at about 10 °C above the temperature of the receiving water. Dis-sipation of the heat depends chiefly on mixing of the hot with cold water. In temperate seas, heated discharges are generally of little consequence, but in tropical seas, where summer temperatures are already near to the thermal death point of many organisms, the increase in temperature can cause substantial loss of life.

2. *Acids and alkalis.* Seawater has a large buffering capacity and the effect of such discharges is extremely localized.

3. *Cyanide.* This comes principally from metallurgical industries. Cyanide rapidly dissociates in seawater and has little effect except in the immediate neighbourhood of the outfall.

Particulates

Inert particulate matter may clog the feeding and respiratory structures of animals, reduce plant photosynthesis by reducing light penetration, and, when it settles on the bottom, smother animals and change the nature of the seabed. Such materials include:

(1) dredging spoil;

(2) powdered ash (fly ash) from coal-fired power stations;

(3) china clay waste;

(4) colliery waste;

(5) clay from gravel extraction. Gravel dredged from the seabed contains a proportion of clay and silt which is

washed out in the course of extraction and settles back on the seabed.

A variety of man-made plastics, which are also inert, reach the sea. These include polystyrene spherules, polythene containers, plastic sheeting, nylon ropes, nets, and other fishing gear.

Conservative wastes

Some materials are not subject to bacterial attack and are not dissipated, but are reactive in various ways with plants and animals, sometimes with harmful effects. The principal categories of such wastes are:

(1) heavy metals (mercury, lead, copper, zinc, and so on);

(2) halogenated hydrocarbons (DDT and other chlorinated hydrocarbon pesticides, polychlorinated biphenyls (PCBs), and so on);

(3) radioactivity.

NATURE OF INPUTS

Although it is convenient to consider materials added to the sea as being in the above categories because of their shared properties and the similarity of the effects they have, it must be remembered that actual inputs to the sea are rarely so simple in their constitution.

1. Power station effluent is chiefly hot water, but also contains some chlorine injected into the water at the intake to discourage marine organisms settling in the cooling system and reducing its efficiency or blocking it. Small amounts of metals are leached from the cooling system and turbines.

2. Urban sewage is principally organic but also contains considerable amounts of metals, oils and greases, detergents, and industrial wastes (since most industrial plants use the same sewers as domestic users), as well as pathogens.

Similar complexity may exist in particular geographical sites. An industrialized estuary has a multiplicity of inputs from surrounding industry, often in great variety, as well as from the urban population. In addition, the estuary receives whatever is carried by the river flowing into it, which may include pesticides and other products of agricultural activity over the area of the entire drainage system.

SOURCES OF INPUTS

Direct outfalls

The most obvious inputs of material to the sea are through pipes discharging directly into it.

1. *Estuaries.* Historically, most ports grew up on estuaries and became centres of population and industry. The urban and industrial wastes were discharged directly into the estuary without treatment. With the growth of population and industry, the major industrial estuaries (Thames, Mersey, Meuse, Scheldt) became foul, stinking, and lifeless towards the end of the last century.

2. *Coastal towns.* Most coastal towns have little industry but, in the past, untreated municipal waste and sewage was discharged directly into the sea by numerous outfalls, sometimes at the high-water mark and rarely far below low water. This resulted in the fouling of local beaches, and many holiday resorts have extended the outfalls further offshore to reduce this or have introduced sewage treatment.

3. *Coastal industry.* With increasing pressure to conserve rivers and lakes for drinking water supplies, new industry with demands for large volumes of water (for example, for cooling or effluent disposal) has tended to be sited on the coast: nuclear power stations in Britain are nearly all sited on the coast.

River inputs

Rivers flow via their estuaries to the sea and

potential pollutants reach rivers over the entire drainage area.

1. Organic wastes are subject to bacterial attack in transit and, depending on the distance of major inputs from the river mouth, the organic load entering the sea from upstream is correspondingly reduced and may be negligible.

2. Pesticides and fertilizers from agriculture and forestry are washed off the land by rain, enter water courses, and eventually reach the sea.

3. Petroleum and oils washed from roads by rain enter the sewage system as a rule, but storm water overflows carry these materials into rivers and to the sea.

All land-based materials washed off by rainfall and entering rivers and streams contribute inputs to the sea, and in aggregate they may form a very large contribution.

Shipping

Ships carry many toxic substances in the course of trade: oil, liquefied natural gas, pesticides, industrial chemicals, and so on. Shipwrecks and other accidents at sea may release these substances, and the very large size of some vessels—crude oil carriers of up to 350 000 tonnes, for instance—means that when accidents do occur, the consequences may be very damaging. Spectacular oil tanker wrecks are an obvious example.

Other shipping is responsible for less dramatic inputs of materials to the sea, both as a result of accidents and in the course of routine operations. A ship does not have to be wrecked and become a total loss for this to happen; in all but the worst accidents much of the cargo is salvaged, but particularly noxious or dangerous materials are frequently carried as deck cargo as a safety precaution for the ship, and these may be lost overboard in severe storms.

In the course of routine operations, ships discharge oily ballast water and bilge water (not always legally) and cargo tank washings; they also discard much litter, of which plastics probably constitute the worst nuisance.

Offshore inputs

A variety of material is dumped at sea in designated dumping grounds.

1. *Dredging spoil.* Shipping channels in estuaries and at the entrances to ports may require frequent dredging to keep them open, and the dredged material is barged out to sea and dumped. This dredging spoil, particularly from industrialized estuaries, may contain appreciable quantities of heavy metals and other contaminants which are then transferred to the dumping grounds.

2. *Sewage sludge.* Sludge from sewage treatment plants is dumped at sea in considerable quantities by Britain; less so by other countries. It has a high organic content but is also contaminated with heavy metals, oils and greases, and many other substances.

3. *Other wastes.* Fly ash from power stations and colliery waste are dumped at sea, as are various dangerous materials such as radioactive waste and ammunition.

4. *Offshore industrial activities.* These result in a variety of inputs or disturbances in the sea. They include offshore oil exploration and extraction; gravel extraction; and in the future, manganese nodule extraction.

Atmospheric inputs

Discharges to the atmosphere are returned to land or the sea in rain or, if particulate, as fall-out. Gaseous wastes dissolve directly in the sea at its surface. Such inputs are, of course, on a regional or even global scale rather than local. There are too many unknown factors for atmospheric inputs to the sea to be estimated precisely, but they are generally supposed to be very large and to represent major contributions.

The total global input of lead to the sea from natural and man-made sources is estimated to be about 400 000 t yr^{-1}. Of this, about half is derived from vehicle exhausts containing leaded-petrol additives which reach the atmosphere and are then rained out.

The total global input of mercury to the sea from volcanic activity and weathering of mercury-bearing ores is at least 5000 t yr^{-1} but may be as high as 25 000 t. A further 5000 t yr^{-1} are added from the use of mercury and mercurial compounds in industry and agriculture. An additional 3000 t are derived from burning fossil fuels, principally coal. Although coal contains only minute amounts of mercury, it is burned in such quantity that, in total, it makes a major contribution to mercury in the sea.

Such global figures are estimates only and cannot be regarded as accurate measurements, but it is clear that materials discharged to the atmosphere can make unexpectedly large contributions to the budget of these contaminants in the sea.

These two examples also reveal that there is often a substantial natural input of materials which, in some circumstances, prove to be pollutants.

DEFINING POLLUTION

The word 'pollution' is commonly used to mean

(1) the environmental damage caused by wastes discharged into the sea;
(2) the occurrence of wastes in the sea;
(3) the wastes themselves.

This is confusing and does not encourage a detailed analysis of the effects of the wastes in the sea. Most pollution scientists use different terms for the wastes ('inputs'), the occurrence of them in the sea ('contamination'), and the damaging effects they have ('pollution'). These distinctions are also recommended by such international advisory bodies as the United Nations Group of Experts on the Scientific Aspects of Marine Pollution (GESAMP) and the International Commission for the Exploration of the Sea (ICES).

Inputs

Some of the wastes reaching the sea (for example, many pesticides, plastics) are man-made and do not otherwise occur in nature, but most of the substances discussed so far exist naturally in the sea:

(1) organic material subject to bacterial degradation;
(2) metals in the run-off from metalliferous deposits;
(3) oil from natural seeps such as those occurring in parts of the Gulf of Mexico, on the Californian coast, and parts of the British coast;
(4) particulate material from coastal erosion—on the north-east coast of England this includes coal from seams exposed on coastal cliffs;
(5) hot water from geothermal springs around the coast of Iceland, in the Galapagos Trench, and elsewhere;
(6) radioactivity on the coasts of Brazil and south-west India.

Because of this variety of natural sources, it is confusing to refer to inputs to the sea as 'pollution'. Are all inputs, natural as well as man-made, to be regarded as polluting? Or are only man-made inputs polluting, even though natural inputs of the same substances may be very much greater?

Contamination

With so many natural inputs to the sea in different parts of the world, the concentrations of substances vary widely from place to place in the marine environment. **Contamination** is caused when a man-made input increases the concentration of a substance in seawater, sediments, or organisms above the natural background level for that area and for the organisms. This locally elevated concentration may, of course, be less than the concentration of the same substances in other areas where there is a large natural input.

The sensitivity of modern chemical analytical techniques now makes it possible to detect extremely low concentrations of many substances in the sea. The apparent spread of contamination in recent years is due more to the improvement in analytical techniques than to any change in the quantity of materials entering the sea.

Many substances change their nature when added to seawater: compounds may dissociate or become ionized; metals may change their valency or form complexes with organic molecules; substances may dissolve in seawater or become adsorbed on to particulate matter and be carried to the seabed. These physico-chemical changes affect the degree to which the added substances are available to marine organisms and the effects they have on them.

For these reasons, a simple measure of the concentration of a substance in the sea is unlikely to reflect the effects it may have.

Pollution

The matter of the greatest concern is, of course, the effects the inputs have in the marine environment. GESAMP defined pollution to reflect this aspect of wastes in the sea. Marine **pollution** is the introduction by man, directly or indirectly, of substances or energy to the marine environment resulting in deleterious effects such as: hazards to human health; hindrance of marine activities, including fishing; impairment of the quality for the use of seawater; and reduction of amenities. The focus is therefore on man-made rather than natural inputs to the sea, and on the damaging effects of wastes.

Practical implications

It can be argued that, with the advance of scientific knowledge, what at one time was believed to be a harmless level of contamination (that *is not* 'polluting') may later be found to cause subtle damage (it *is* 'polluting'). As a precaution against such unpleasant surprises, it would be prudent not to discharge any wastes into the sea.

Even if this policy is adopted, it does not remove the need to distinguish between inputs, contamination, and pollution. Many inputs to the sea are not deliberate but are derived from atmospheric fall-out, run-off from the land, or accidental spillages. In addition, although the discovery of high con-

centrations of a substance may provide a warning signal, unless it is the result of human activities and is damaging, it does not constitute pollution. It should be noted, too, that even if laboratory studies show that a contaminant is toxic to particular marine organisms, this may provide another—possibly stronger—warning signal, but does not necessarily prove that the contaminant has harmful effects in the natural environment.

From a strictly biological point of view, even if toxic contaminants do cause the death of some plants and animals in the natural environment, this is usually of less consequence than if the deaths result in a change in the population as a whole. Most marine animals reproduce on a colossal scale and the overwhelming majority of offspring die prematurely in the natural course of events. Mortality from toxic contamination may be insignificant compared to these natural losses and may have no effect on the population.

It follows from these considerations that, whereas there can be no doubt about the existence of pollution when the damage is severe, there is often considerable difficulty in identifying damage from low levels of contamination caused by diffuse, or sometimes unknown, inputs. Pollution is by no means as clear cut as it might at first appear.

PRIORITIES

It will be noticed that in defining pollution, most examples quoted of its deleterious effects relate to human interests in the sea. Pollution is, in practical terms, an example of one set of human interests:

(1) the use of leaded petrol in cars;

(2) burning coal;

(3) transporting oil;

(4) generating electricity;

(5) disposing of waste products

coming into conflict with other human interests:

(1) human health;

(2) commercial fisheries;

(3) amenities, tourism, recreation, and aesthetic values;

(4) scientific interest.

Generally, we would consider transport, the provision of electricity, and many vital industrial processes to be as important as preserving a fish supply or providing tourist facilities. So, we are really concerned with striking the most favourable balance (for humans) between these conflicting interests—in other words, we are concerned with priorities.

As a rule, a threat to human health is given the highest priority for abatement or control. There are too many examples to the contrary to claim that it is given overriding priority over all other considerations, but it is certainly an issue which is given the greatest importance.

When only two interests conflict and both are commercial, it is fairly easy to decide the priorities. When the oil tanker *Torrey Canyon* was wrecked off the Cornish coast in 1967 and deposited a large amount of oil on the beaches, a decision had to be taken whether or not to treat the spilled oil with chemical dispersants which, at that time, were very toxic to marine organisms: their use would have meant a serious risk of great damage to the local fishing industry, but to leave the oil untreated would have been to the great detriment of the tourist industry. The fishing industry was then worth £6 million per year and the tourist industry £60 million per year, so on that basis the decision was taken to clean the beaches. Happily, in the event fears about damage to the fishing industry proved to be unfounded and it was not noticeably affected.

Generally, however, the issues are not so clear cut and the decisions are more difficult. A more sophisticated approach is needed and, for this, a great deal of information is required:

1. What is the level of contamination in the area we are interested in?

2. What form does it take?

3. Where does it come from?

4. What happens to it?

5. What does it do to the plants and animals there?

6. If plants and animals are affected, does it matter?

7. To whom does it matter? What other interest is affected?

8. How much does it matter to them?

9. If it does matter, what can we do about it?

10. What do we do with the polluting material if it is not put in the sea?

11. Would the alternative be better or worse than putting it in the sea?

12. How much would it cost?

COST OF POLLUTION ABATEMENT

A number of these questions involve matters which are economic or political and are outside the realm of pure science. Human activities inevitably produce wastes that must be disposed of somewhere. While it may be possible to avoid creating some wastes by ceasing to use a product or by recycling, many wastes (such as human sewage) are unavoidable.

Waste disposal, however it is managed, has both an environmental cost and a financial cost. While we wish to minimize the environmental costs, the financial costs cannot be ignored because, depending on the environmental standards demanded, they can very rapidly escalate.

The cost of designing a new coal-fired power station to remove particulates from the waste gas is as follows:

(1) 90 per cent extraction of particulates adds 10 per cent to the capital cost;

(2) 95 per cent extraction adds 20–30 per cent;

(3) 99 per cent extraction doubles the capital cost of the power station.

A sugar beet processing plant handling 2700 tonnes per day produces an effluent high in organic wastes. The organic content is measured as BOD (biochemical oxygen demand, see p. 9) and this can be reduced by suitable treatment, but at a cost:

(1) 30 per cent reduction in BOD costs less than $0.50\,kg^{-1}$;

(2) 65 per cent reduction in BOD costs about $10\,kg^{-1}$;

(3) 95 per cent reduction in BOD costs about $30\,kg^{-1}$.

A similar exponential relationship between effluent standards and treatment costs exists for most wastes (Fig. 1.1): at the margin, costs can easily double for a trivial improvement in environmental quality.

A policy decision is required about where on this exponential curve a standard should be set. Scientists can probably predict fairly well the effect on the local environment of a particular level of discharge, but they cannot decide the appropriate trade-off between environmental cost and the financial cost of disposing of the waste. That must be a political decision, made by taking account of other matters such as: the value placed on the local environment, public perceptions of risks, and social costs if a factory is forced out of business by high effluent treatment costs.

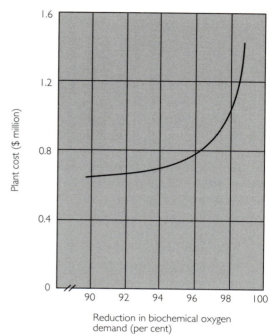

Fig. 1.1 Capital cost of treating an effluent containing organic and inorganic industrial waste. (*Published with the permission of the Controller of Her Majesty's Stationery Office.*)

Nevertheless, it is important that the scientist has as clear a view as possible of the environmental consequences of waste discharges if wise political decisions are to be made.

2

Oxygen-Demanding Wastes

The amount of oxygen dissolved in the water has a profound effect on the plants and animals living in it. Wastes which directly or indirectly affect the oxygen concentration are therefore of great importance.

Oxygen Demand

By far the greatest volume of waste discharged to watercourses, estuaries, and the sea is sewage which is primarily organic in nature and subject to bacterial decay. Bacterial degradation results in the oxidation of organic molecules to stable inorganic compounds. Aerobic bacteria make use of oxygen dissolved in the water to achieve this.

$$C_6H_{12}O_6 + 6O_2 \rightarrow 6H_2O + 6CO_2$$
(glucose) (oxygen) (water)   (carbon dioxide)

As a result of this bacterial activity, the oxygen concentration in the water is reduced but this is compensated for by the uptake of atmospheric oxygen. However, oxygen diffuses only slowly through water and there is a time-lag before the oxygen used by bacterial activity is replenished. When the oxygen concentration falls below about 1.5 mg l^{-1}, the rate of aerobic oxidation is reduced. Anaerobic bacteria can oxidize organic molecules without the use of oxygen, but the end-products include compounds such as H_2S (hydrogen sulphide), NH_3 (ammonia), and CH_4 (methane), which are toxic to many organisms, and this process is much slower than aerobic degradation. There is therefore the likelihood of waste accumulation.

Some inorganic wastes become oxidized in water without the intervention of bacteria, and these, too, deplete the water of oxygen.

When planning the discharge of any waste to water, it is important to know its total oxygen demand to avoid undesirable environmental consequences. The chemical composition of nearly all organic wastes is extremely complicated and different constituents require different amounts of oxygen to achieve complete oxidation. It is impracticable to analyse a type of waste to discover its exact mix so other methods are used to measure the effluent load it represents by measuring the overall oxygen demand for complete degradation.

Chemical oxygen demand (COD)

This is measured by adding an oxidant, such as potassium permanganate ($KMNO_4$) or potassium dichromate ($K_2Cr_2O_7$), with sulphuric acid (H_2SO_4) to a sample of the effluent. The sample is then titrated after a standard interval to determine the amount of oxidant remaining. From this, the total amount of oxidizable material can be calculated.

Biochemical oxygen demand (BOD)

For organic wastes, it is more usual to measure the oxygen concentration in a sample before and after bacterial digestion for a standard time (for example 3 days or 5 days), recorded as BOD_3 or BOD_5. It may be necessary to add bacteria and nitrate if these are deficient in the initial sample. This gives a direct measure of the oxygen utilization in bacterial degradation of the sample.

THE DILUTION FACTOR

If water is saturated with oxygen, there is sufficient to oxidize a BOD of about 8.0–8.5 mg l^{-1}, but the BOD of organic effluents is usually much greater than this. Urban sewage commonly has a BOD of 500 mg l^{-1}; spilled beer has a BOD of 70 000 mg l^{-1}. It is therefore necessary to find some way of diluting the effluent to achieve a BOD of about 8 mg l^{-1}, hence the old adage of sanitary engineers: 'The solution to pollution is dilution'.

The common practice is to discharge the effluent into a large volume of water, preferably a river or the sea, where water movement produces mixing and achieves the necessary dilution. If the effluent is discharged into a river (Fig. 2.1), it is swept downstream and

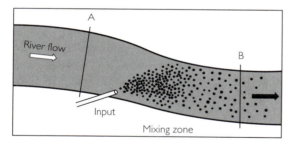

Fig. 2.1 Discharge of an organic effluent into a river and its mixing zone. Position A upstream of the input and position B downstream of the mixing zone are explained in the text.

mixes with the river water in a **mixing zone**. The concentration of effluent is high near the outfall but progressively reduced downstream. The dilution achieved depends upon the rate of flow of the river, its organic load, and the rate of input of the effluent and its oxygen demand.

River flow   8 m^3 s^{-1} with BOD   2 mg l^{-1}
Effluent input   1 m^3 s^{-1} with BOD 20 mg l^{-1}

$$\text{BOD after mixing} = \frac{\text{total BOD}}{\text{total volume}}$$

$$= \frac{(8 \times 2) + (1 \times 20)}{8 + 1} = 4$$

This achieves the necessary dilution.

OXYGEN BUDGET

Bacterial activity consumes oxygen. Figure 2.2 shows the effect of an organic input on the dissolved oxygen in the water, either with time following a single input to a static body of water, or with distance downstream from a continuous input to flowing water.

As the bacteria multiply, oxygen consumption increases to a peak and then declines as the organic matter is oxidized. This extracts oxygen from the water column and when the dissolved oxygen falls below saturation, oxygen is taken up from the atmosphere; the greater the oxygen depletion of the water, the greater the rate of oxygen uptake. Uptake lags behind utilization because of the time taken

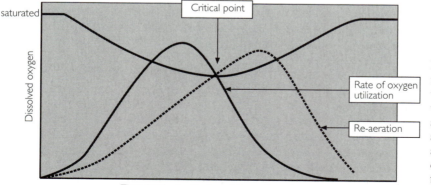

Fig. 2.2 Oxygen budget for a river or body of water receiving organic wastes. The horizontal axis represents time for a single input of waste to a static body of water, or distance downstream from a continuous input to a river.

for oxygen diffusion through the water. At a **critical point**, the rate of reaeration overtakes the rate of oxygen utilization and this corresponds with the minimum concentration of dissolved oxygen in the water.

Thus, the effect of an organic input is to produce an **oxygen sag**, corresponding to the dissolved oxygen in the receiving water, less bacterial respiration, plus reaeration. The oxygen sag may be trivial, but if the receiving waters are seriously overloaded with organic material it may be sufficient to produce anoxic conditions.

FLUCTUATIONS IN DISSOLVED OXYGEN

In practice, the situation is more complicated than this, because the organic input leads to enrichment and the growth of phytoplankton which, because of its photosynthetic activity, provides a supplementary source of oxygen. This shows a **diurnal fluctuation** in response to the diurnal light regime. Thus, if dissolved oxygen is measured upstream and downstream of an organic input (as shown by A and B in Fig. 2.1), with a correction for the flow rate so that measurements are made in the same body of water, greater fluctuations are detected downstream than upstream (see Fig. 2.3). This is because the dissolved oxygen downstream consists of:

incoming dissolved oxygen (at A)
+ photosynthesis by phytoplankton
+ photosynthesis by fixed plants
− respiration of plankton
− respiration of benthos
− oxygen used by bacteria in oxidizing
  organic matter
± oxygen exchange at the water surface.

Upstream of the outfall, there is diurnal fluctuation of dissolved oxygen in the water because of the photosynthetic activity of the plants. The discrepancy between this and the dissolved oxygen downstream of the outfall is the result of the increased plant biomass and its photosynthetic activity.

There is also a **seasonal variation** in dissolved oxygen, to which a number of factors contribute. Because of the greater day-length and light intensity, diurnal fluctuations in photosynthetic activity are greater in summer than in winter, but in summer the higher water temperature results in greater bacterial activity and a higher oxygen demand. Furthermore, river flow—and hence dilution—is least in summer and greatest in winter and spring. The high water temperature combined with low flow rate in summer is most likely to cause problems by increasing the oxygen sag.

ESTUARIES

The situation in tidal waters is much more complicated than in rivers.

Settlement
The previous calculations (p. 10) assume that all the discharge is in solution or suspension, but in practice the effluent often contains solids which settle to the bottom, and it is important to know the settlement characteristics of the discharge if the dilution factor is to be properly calculated. The rate of settlement of particles in water depends on their size and density, and the viscosity and speed of the water; it is governed by Stokes's Law and Newton's Law. However, particles tend to flocculate into larger aggregations and

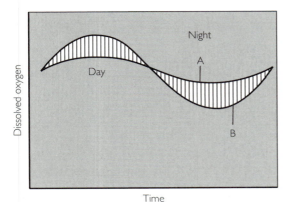

Fig. 2.3 Diurnal fluctuations of dissolved oxygen in a river: A, upstream of an input of organic waste; B, downstream of the mixing zone.

Stokes's and Newton's Laws do not apply in these circumstances.

In estuaries, the change of pH and redox potential as freshwater encounters the sea causes intense flocculation of clay and other particles, with increased adsorption of metals and other materials on the flocculates. While sedimentation takes place in all but the swiftest rivers, estuaries are subject to particularly heavy sedimentation, leading to the development of extensive mud-flats containing much of the organic material, metals, pesticides, and so on from the water column.

No two situations are the same and a detailed study of the settlement characteristics of an effluent in the receiving waters is needed before it can be calculated how much, where, and at what rate waste can be discharged, to reduce the BOD to within the capacity of the receiving waters.

Residence time

Whereas rivers flow in one direction, there is a tidal ebb and flow in estuaries and the net seaward flow over the complete tidal cycle may be small. Mean figures for the Thames at London are:

> high water to low water:
> downstream flow 15 km

> low water to high water:
> upstream flow 13 km

giving a net seaward flow of only 2 km per tidal cycle. Towards the seaward end of the estuary, the seaward flow becomes very slow indeed (Fig. 2.4).

Thus, a body of water (or effluent) entering an estuary moves only slowly through it and has a long **residence time** in it. It follows that the dilution capacity of the estuary is correspondingly reduced. The oxygen sag is more pronounced and may create an **anoxic zone** from which most life disappears, except for anaerobic bacteria, fungi and yeasts, and some protozoa. It becomes foul-smelling from the products of anaerobic oxidation, and migratory fish such as salmon—or more particularly their young, which require a min-

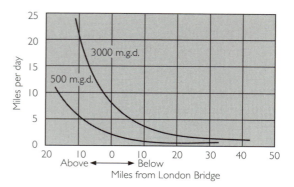

Fig. 2.4 Net seaward flow of water in the Thames estuary at two rates of river flow in million gallons per day. (*Published with the permission of the Controller of Her Majesty's Stationery Office.*)

imum oxygen concentration of 4 mg l^{-1}— cannot pass through this block.

Stratification

Depending on their shape, some estuaries, such as the Tees, show **stratification**, with the denser seawater flowing upstream in a wedge and the river water flowing out over it (Fig. 2.5). Other estuaries, such as the Thames, are not stratified. If an effluent is discharged into the bottom water layer of a stratified estuary and has neutral buoyancy, it may actually move upstream until it reaches the end of the wedge and comes under the influence of the seaward moving upper layers of water. If it has negative buoyancy, it may never travel seaward at all.

Mixing zone

In rivers, there is a defined mixing zone downstream of the point of effluent discharge (Fig. 2.1). In estuaries, the mixing zone is downstream of the point of discharge during an ebbing tide; but on a rising tide the mixing zone may swing upstream from the point of discharge and, as a result, a considerable stretch of the estuary comes under the influence of the effluent. Particularly in stratified estuaries where different effluents behave differently, the whole estuary constitutes the mixing zone.

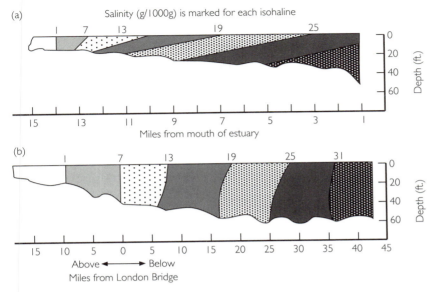

(a)

Salinity (g/1000g) is marked for each isohaline

Miles from mouth of estuary

(b)

Above ◄──────► Below
Miles from London Bridge

Fig. 2.5 Vertical distribution of salinity at high water of spring tides in (*a*) the Tees and (*b*) the Thames estuaries. (*Published with the permission of the Controller of Her Majesty's Stationery Office.*)

SEWAGE TREATMENT

River water with a BOD_5 of less than 2 mg l^{-1} can be regarded as unpolluted; with a BOD_5 greater than 10 mg l^{-1} it is grossly polluted. Between these extremes, various standards may be set, depending on the purpose for which the river water is required. Although they have not been adopted in all member states, the EC guidelines on water quality give an appropriate impression of the BOD standards required. Water for salmon and trout should have a BOD_5 less than 3 mg l^{-1}, for cyprinid fish less than 6 mg l^{-1}. Three standards for drinking water supplies have a BOD_5 less than 3 mg l^{-1}, less than 5 mg l^{-1}, or less than 7 mg l^{-1}.

A common dilution factor in rivers is 8:1. To achieve a BOD_5 of 4 mg l^{-1}, which might be acceptable for many purposes, the BOD_5 should not exceed about 20 mg l^{-1} (see example on p. 10). If more waste has to be disposed of than the receiving waters can accept, and it is impossible to get the required dilution, it is necessary to introduce sewage treatment in which favourable conditions are provided for accelerated bacterial degradation of part of the organic material. This will reduce the BOD of the final product before it is discharged to the receiving waters. Various

stages of treatment may be used, depending on the quality of effluent that is required (Fig. 2.6).

Primary treatment

The sewage is screened to remove large solids, it is comminuted or macerated to reduce it to a slurry, grit is extracted, and the remainder is placed in settlement tanks. The supernatant liquid, which still has a high BOD, and a certain proportion of suspended solids is discharged to the receiving waters, and the sludge which has settled out is disposed of elsewhere.

Secondary treatment

If a greater reduction of BOD is required, the liquid is subjected to biological treatment. It may be filtered through beds of 4–15 cm rocks or coke, providing a large surface area for bacteria which degrade organic matter in the water as it percolates through the bed. Material settled from the outflow may be disposed of and the supernatant liquid discharged to the receiving waters.

The sludge from the first stages of digestion gains a very high bacterial count (**activated sludge**). This may be retained, and air blown through a suspension of the effluent and

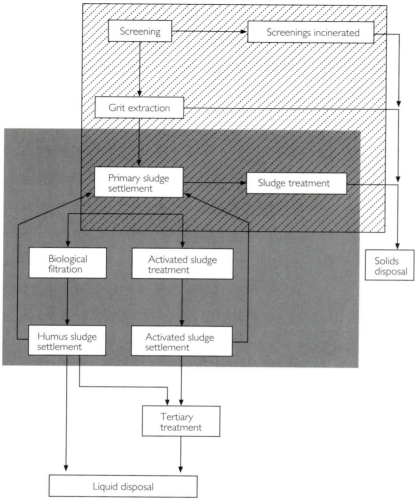

Fig. 2.6 Scheme of primary (cross-hatched) and secondary (shaded) treatment of sewage.

activated sludge; very high rates of bacterial activity can be achieved by this method.

Tertiary treatment

If a high quality of effluent is required, it may be necessary to treat it further. The liquid derived from secondary treatment may be retained in sedimentation ponds or passed through sand or earth filters to remove suspended solids. Nitrates may be removed by algal growth in retaining ponds and phosphates can be removed by electrolytic methods.

The provision of sewage treatment plants is costly: a modern treatment plant giving secondary treatment to sewage for a population of 12 500 people costs over £4 million to build; providing the sewers, and so on costs 2–3 times as much again; and the annual running costs are about 5 per cent of the capital costs. In view of this, no more sewage is treated than necessary, and advantage is taken of the natural capacity of receiving waters to degrade wastes, so far as this can be done safely.

DISPOSAL OF SEWAGE AND SLUDGE

Human sewage may be disposed of in cess pits

or septic tanks or, in some cases (notably the city of Brussels), it is discharged, untreated, into rivers. In countries where the great majority of the population is connected to the sewerage system, the wastes are pumped to plants where they may receive some treatment before discharge.

Nearly all coastal communities discharge their sewage into the sea via pipelines. Where the pipeline is short or badly sited, this practice results in beaches becoming contaminated with sewage. The introduction of an EC directive on bacterial standards for bathing beaches has led to coastal resorts constructing longer pipelines to ensure that sewage does not reach the beaches, or introducing primary or secondary treatment of the sewage before discharge. In some cases, the effluent is disinfected with chlorine or by other means; though this also has its dangers if organochlorine compounds are formed as a result of this treatment.

Sewage treatment produces large quantities of settled solids which are generally not discharged with the liquid effluent, but are disposed of in some other way. In some respects this may be regarded as transferring the pollution from one place to another, but the objective is to distribute the organic load in such a way that it does least harm and possibly some good.

Ideally, organic wastes should be returned to land to fertilize crops. Some raw sewage and pig waste is sprayed on crops as a slurry and, of course, human sewage is widely used as an agricultural fertilizer in China and elsewhere in the Third World. However, the use of untreated sewage and animal manure on crops causes an objectionable smell and a nuisance from flies, as well as creating a small public health risk (see p. 25). Sewage sludge is more suitable for use as an agricultural fertilizer, and nearly half the sewage sludge produced in Britain is used in this way.

There is probably little scope for increasing the proportion of sewage sludge used in agriculture: sewage sludge is not suitable for use on all types of soil; it is difficult to adjust the rate of application to provide an appropriate nitrogen and phosphorus level; and its regular use on agricultural land results in an accumulation of metals in the soil. One of the highest soil concentrations of cadmium in England was found in a 400-year-old garden far from any industry or centre of population, but which had been fertilized with human wastes for several centuries. Urban sewage contains industrial wastes and is contaminated with metals, detergents, and other toxic material which make it unsuitable for agricultural use. Above all, there is simply too much sewage for it all to be returned to agricultural land. One thousand people produce, on average, 25 tonnes of solid waste per year and intensive animal rearing adds to the problem of waste disposal. The Netherlands has recently had to limit the size of its pig industry because of the difficulty of disposing of such a quantity of pig waste.

A considerable quantity of sewage sludge is therefore disposed of on land tips, or incinerated. Until recently, cities with reasonable access dumped sewage sludge at sea, but increasing concern about the impact of sludge dumping in the marine environment (see p. 18) has resulted in the progressive reduction of this practice. This creates new problems.

Suitable landfill sites are a long way from major coastal conurbations (for example New York and London) and transport of sludge to dump sites is difficult. Incineration of human and animal waste can be harnessed to provide thermal energy which may be put to useful purposes, as is practised in Denmark and to some degree in other countries, but incinerators are usually objected to by the population near any proposed site.

Sewage sludge has a high water content and usually has an unpleasant smell. To reduce transport costs, it is therefore **dewatered** and commonly subjected to further treatment to produce a more acceptable material for tipping or incineration. The sludge is retained in shallow drying pools where, by drainage and evaporation, the water content is reduced from 90 per cent to 45–60 per cent. Because of the increased time that the sewage must remain at the treatment plant, a correspondingly greater area of land has to be devoted to this activity. The sludge may then be exposed to further bacterial action under heat and this treatment results in the production of combustible gases which can be used as a fuel source in the sewage treatment works. The sewage, if properly treated, acquires a peaty or tarry smell and is then more acceptable for dumping on tips.

CONSEQUENCES OF ORGANIC DISCHARGES TO ESTUARIES

The most serious problems are encountered in industrialized and urbanized estuaries. Two case histories illustrate contrasting difficulties and the measures adopted to reduce them.

Thames estuary

The Thames was a major salmon river until about 1820. Sewage from London was either carried to fields in the surrounding countryside and used as fertilizer or simply left to rot in cess pits or the streets. The population was largely dependent on wells for its drinking water; these were frequently contaminated and typhoid was common and cholera epidemics occurred at intervals. Then came an improved piped water supply, the water closet was introduced, and construction of the main system of sewers was begun, discharging the wastes into the Thames. As a result of these public health measures, disease was reduced, but the river was progressively overloaded and by the 1850s had become foul smelling and devoid of fish.

A reduction in the number of separate outfalls and increase in sewage treatment around the turn of the century greatly improved the condition of the river, but the growth of London's population outran the capacity of the sewage treatment system; by the 1950s, about 20 miles of the Thames through London was anoxic in dry weather (Fig. 2.7). Improvements and expansion of sewage treatment works around 1960 resulted, within 10 years, in a dramatic increase in the numbers and variety of benthic invertebrates, marine

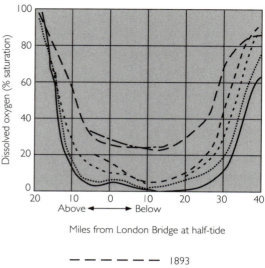

Miles from London Bridge at half-tide

– – – – – –	1893
–·–·–·–	1900–1905
– – – – – –	1920–1929
·–·–·–·–	1930–1939
··············	1940–1949
——————	1950–1959

Fig. 2.7 Oxygen sag in the tidal river Thames at various times since 1893. (*Published with the permission of the Controller of Her Majesty's Stationery Office.*)

and estuarine fish, and shore birds and waders feeding on the mud-flats in winter.

Mersey estuary

The Mersey is probably the most heavily polluted estuary in Britain and cleaning it up presents an intractible problem.

The river Mersey and its tributaries pass through Greater Manchester, Oldham, Bolton, Rochdale, Stockport, and other heavily industrialized areas with a total population of 3–4 million. Although sewage and industrial wastes are treated before they are discharged into the rivers, the sewage treatment capacity is limited and sewage overflows into rivers at numerous points in wet weather. Thus, unlike the industrial estuaries of the Tyne and Tees, and to a degree the Thames, which have unpolluted rivers entering them, the Mersey receives a very heavy effluent load upstream.

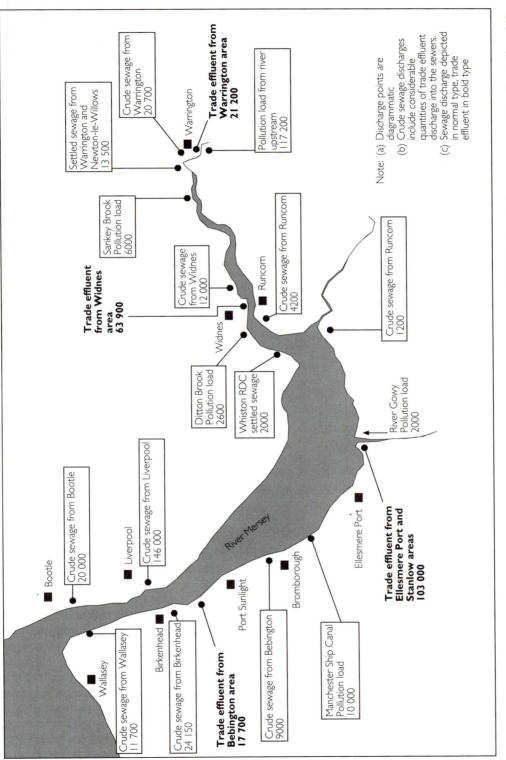

Fig. 2.8 Sewage and industrial waste inputs (in lb BOD per day) to the Mersey estuary. (*Published with the permission of the Controller of Her Majesty's Stationery Office.*)

In addition, the Manchester Ship Canal, constructed in 1893, carries much of the river water, bypassing the upper and middle parts of the estuary and entering it at Eastham. The canal gates to the lower estuary are opened only intermittently to maintain the water level, and there is therefore a very small fresh-water input to the estuary.

From Warrington, at the head of the estuary, downstream, the numerous inputs of sewage and industrial wastes are mostly untreated (Fig. 2.8). They discharge into the estuary or the ship canal. Liverpool alone has 26 separate outfalls into the lower estuary. There is thus a heavy effluent load in the estuary, which depends on tidal flushing for its removal. At high tide, seawater tends to push back water in the ship canal which over-flows, with its effluent load, into the middle estuary.

The effects, predictably, are a low dissolved oxygen level in the upper estuary which becomes anoxic in summer, and heavy metal contamination of the mud-banks throughout most of the estuary. The main problem in the lower estuary is the presence of visible solid wastes, much of which come ashore on amenity beaches at New Brighton, Crosby, and Formby, outside the mouth of the estuary.

It would obviously be a long and costly task to bring the Mersey estuary to the healthy condition of the Thames, but some improvement has been made in recent years. The first priority is to improve the water quality of the rivers entering the estuary and to reduce visible pollution in the lower estuary and on amenity beaches. Increased sewage treatment capacity has already reduced the incidence of anoxic conditions in the upper estuary down-stream of Warrington.

There has also been a substantial reduction of industrial effluent discharge directly into the estuary, from a BOD_5 of 104.5 t day^{-1} in 1972, to 71.6 t day^{-1} in 1979, and 33.6 t day^{-1} in 1986. Some of the improvement has been due to a decline in industrial activity, but much is due to the remedial measures that have been taken. A new sewer is under con-struction to intercept the 26 separate outfalls

from Liverpool and conduct the sewage to a new treatment plant before discharge to the lower estuary. This will do much to remove the visible pollution of neighbouring beaches. These measures attack the most conspicuous problems, but very much more will need to be done to restore the Mersey and its estuary to an unpolluted condition.

CONSEQUENCES OF SLUDGE DUMPING AT SEA

Although in some cases this practice has now stopped or is being phased out, it has been common to dump sewage sludge at sea where it was practicable to do so. The effect of this input depends on the hydrography at the disposal sites, and it is instructive to compare the different experiences at some of these areas.

Firth of Clyde

Sewage sludge has been dumped in the Firth of Clyde, in the west of Scotland, for over a century. Since 1974, 1.5 m t yr^{-1} has been deposited at a site off Garroch Head at the south end of the island of Bute (Fig. 2.9a). (Until 1974, a different site 6 km closer to Garroch Head was used.) The water depth is 70–80 m and bottom currents are weak, with the result that the wastes accumulate. There is sufficient water exchange that, despite the organic enrichment of the seabed, it does not become depleted of oxygen.

The bottom fauna is modified, with the polychaete *Capitella* and other species characteristic of organically enriched sub-strata dominant at the centre of the dump site. Surrounding this heavily impacted area, there is a transition to a normal benthic fauna within 3–5 km (Fig. 2.10). Metals accumulate in the sediments at the dump site and, although the fauna at the old dump site rapidly returned to normal, the sediments remain contaminated with metals (Figs 2.9b, c, d) Fish and shellfish in the area do not have increased metal concentrations.

There is no evidence that phytoplankton

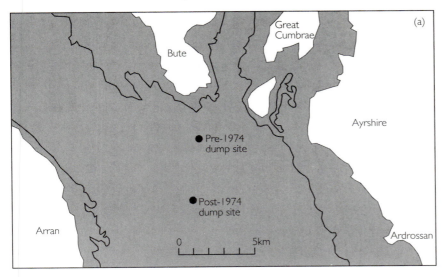

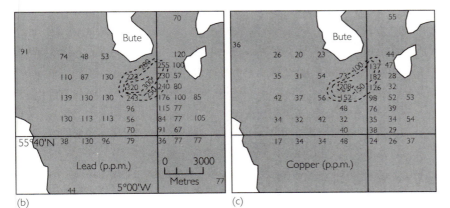

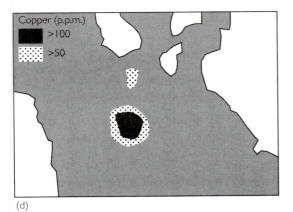

Fig. 2.9 Sewage sludge dumping in the Firth of Clyde: (*a*) position of the dump sites before and after 1974; (*b*) lead concentration in sediments in 1970; (*c*) copper concentration in sediments in 1970; (*d*) copper concentration in sediments in 1980.

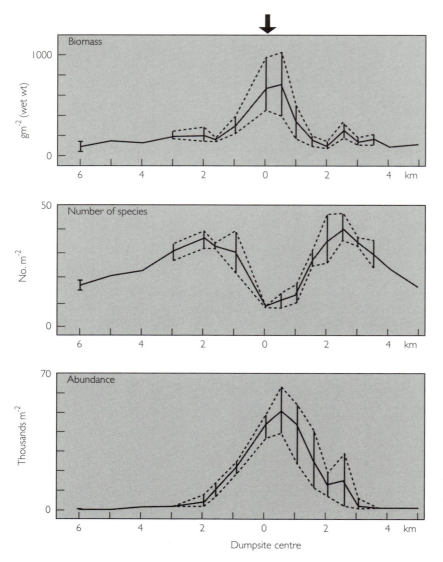

Fig. 2.10 Sewage sludge dumping in the Firth of Clyde: mean and range of (*a*) biomass; (*b*) numbers of species; (*c*) total numbers of individuals in the benthos during the period 1979–85.

production is increased by sludge dumping at the site.

The Garroch Head disposal site is a 'sacrificial' dumping ground and is comparable to a waste tip on land.

Thames estuary

About 5 m t yr^{-1} of sewage sludge has been dumped in the outer Thames estuary for the last century (Fig. 2.11). The disposal grounds are in an area with very mobile bottom sediments and fast bottom currents which disperse and dilute the wastes. The site is evidently used to its full capacity, however. Sediments throughout the area contain particles originating in the sludge and have increased concentrations of organic carbon, and an accumulation of heavy metals in sediments has been identified in one or two small areas. Fish and shellfish in the area had increased concentrations of mercury in the 1970s, but this had fallen to background levels by 1980, following a reduction in the amount of mercury in the sludge.

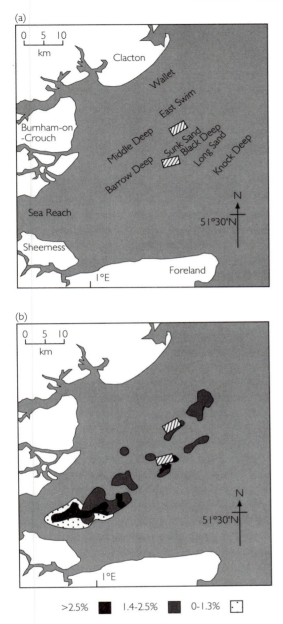

Fig. 2.11 Sewage sludge dumping in the outer Thames estuary: (*a*) location of dumping ground, (*b*) organic carbon content of sediments. (*Pergamon Press.*)

Sediments in the outer Thames estuary are very heterogeneous and support a variety of different assemblages of fauna; because the bottom sediments are mobile, many of these

assemblages are probably temporary. This may account for the fact that in some surveys an enhanced fauna has been found in the area, and in others an impoverished fauna. There is no clear evidence that sludge dumping has had any influence on the local fauna.

New York Bight

Sewage sludge from New York's twenty sewage treatment plants has been dumped at a site in New York Bight since 1924; until recently, 10 m t yr^{-1} were disposed of in this way. The impact is intermediate between that at the sacrificial ground in the Firth of Clyde and that in the dispersive grounds in the Thames estuary. There is organic enrichment and accumulation of metals with faunistic impoverishment in the centre of the dumping area, but despite the very much greater volume of waste dumped, the concentrations of metals reached are only half those found in the Clyde. There is some evidence of the spread of the contaminated area, and con-taminated sediments are carried into the Hudson Trench (Fig. 2.12).

German Bight

A disposal site in the German Bight was in use from 1960 to 1980 and received about 250 000 t yr^{-1} of sewage sludge from Hamburg. Large particles of sludge remained at the disposal site, but finer particles were widely distributed by tidal currents. Except for a thin surface layer, bottom sediments in the dumping area were anaerobic, producing hydrogen sulphide, but similar anaerobic sediments occur naturally in other parts of the German Bight.

It has been difficult to study the effects of sludge dumping because the area also receives the heavily contaminated estuarine waters from the rivers Elbe and Weser, and metal concentrations in sediments at the dump site were no different from those else-where. The dumping site is also in a transition zone between two sediment types, one dominated by the bivalve *Abra alba*, the other by *Nucula turgida*. When studies began in 1970, and for a few years afterwards, areas

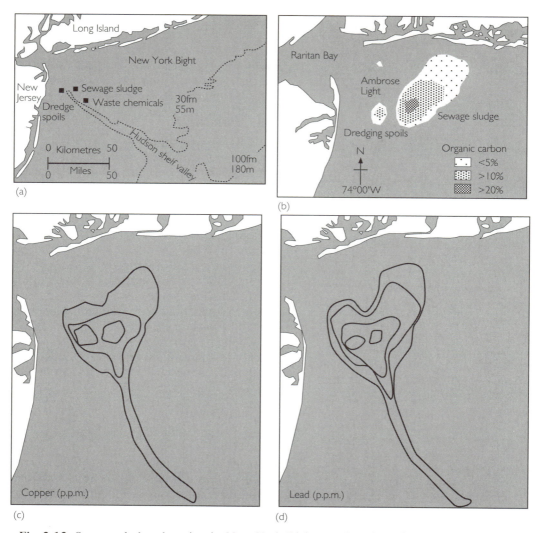

Fig. 2.12 Sewage sludge dumping in New York Bight: (*a*) location of dumping ground, (*b*) organic carbon content of sediments, (*c*) copper, and (*d*) lead in sediments. (*Pergamon Press.*)

receiving sludge, as shown by the presence of tomato seeds, supported a dense population of the *Abra alba* with a typical associated fauna (Fig. 2.13). *Nucula* remained abundant in areas where there was no evidence of sewage sludge. The *Abra* population crashed, and in 1977 and 1978 the benthos in areas where it had been abundant then had a depleted fauna, with a lower biomass and lower species diversity than surrounding areas. Large population fluctuations from natural causes are common and this change is

more likely to have been a reflection of them than to have been related to sludge dumping.

The German Bight was regarded as unsuitable to receive sewage sludge and dumping stopped after 1980. The area is susceptible to periods of oxygen depletion in bottom waters when a thermocline develops (see p. 24). It already receives a large input of plant nutrients from rivers, and the addition of organic matter in the form of sewage sludge is estimated to have contributed 10 per cent to the depletion of oxygen concentrations.

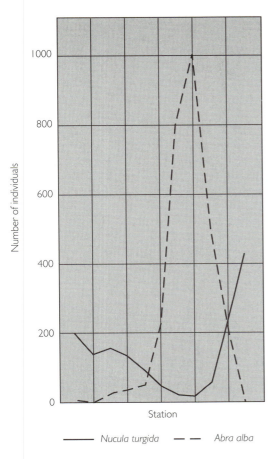

Fig. 2.13 Cross-section of the German Bight sludge dumping ground, showing numbers of *Abra alba* and *Nucula turgida* in 1974. (*Pergamon Press.*)

ENRICHMENT AND EUTROPHICATION

Enrichment

Many of the wastes entering the sea are plant nutrients. Decaying organic matter, and nitrates and phosphates in sewage, enhance plant growth. Another important source of plant nutrients is agricultural fertilizers in run-off from areas of intensive farming. Inputs of plant nutrients to the sea enhance the growth of phytoplankton and fixed plants, and this **enrichment** benefits a number of food chains, as does the increase in bacterial populations stimulated by an organic input. A moderate input of nutrients has the same effect in the sea as adding fertilizers or manure to a garden or farm land. The rich inshore benthic fauna and the important fisheries of the Firth of Clyde may reflect the input of organic wastes (sewage) and agricultural fertilizers in run-off from the surrounding catchment area.

Eutrophication

Over-fertilization of the sea, as on land, has the undesirable effect of altering the structure of communities, sometimes disastrously. **Eutrophication** is a familiar problem in freshwaters: tropical and subtropical rivers and lakes become choked with water hyacinth; eutrophication of parts of the Great Lakes of North America caused population changes in commercial fish stocks. In the sea, a common sign of sewage pollution on a beach is the growth of the green algae *Enteromorpha* and *Ulva*, but until recently, more widespread eutrophication was not regarded as a particularly serious problem in the sea except in one or two areas of restricted water exchange. It has now become clear that this is not so.

Red tides

A large input of plant nutrients often results in the development of **red tides**. These are phytoplankton blooms of such intensity that the sea is discoloured (not always red; the blooms may also be white, yellow, or brown). Many animals, including important commercial fish species, are killed or excluded from the area, either because of clogging of the gills or other structures, or because of the toxic properties of the phytoplanktonic organisms.

In 1981 and 1985 high nitrogen levels were measured in Danish waters during phytoplankton blooms which resulted in mass kills of fish and invertebrates. Blooms of *Phaeocystis* regularly affect the coastlines of northern France, Belgium, the Netherlands, and Great Britain. *Phaeocystis* forms an unpleasant foam which, when stranded on the shore, is often mistaken for sewage pollution.

Whatever biological effects the *Phaeocystis* bloom in the upper Adriatic may have had during summer 1990, it had a devastating economic effect on resorts in the affected area, which were deserted by the tourists.

In May 1988, an exceptionally large and persistent bloom of *Chrysochromulina* in the Skagerrak and North Sea caused serious damage to coastal fish farms in Sweden and southern Norway. This species had not formerly been known to form blooms. While nutrient levels have increased in the Skagerrak in recent years, on this occasion physical and climatic factors appear to have been a more important influence.

Following the damaging phytoplankton blooms in the early 1980s, Denmark introduced controls on the use of fertilizers in agriculture and on the development of coastal fish farms. Germany and the Netherlands have also introduced controls.

Oxygen depletion

A second damaging effect of over-enrichment is that the dense blooms of phytoplankton result in very high oxygen concentrations in surface water because of increased photosynthesis. The abundance of decaying plant material falling to the seabed severely reduces the oxygen concentration in bottom waters and most benthic animals are killed or excluded from the area.

In prolonged, warm, calm weather, a **thermocline** may develop separating cold, dense bottom water from the warm surface layer and from atmospheric oxygen. Bacterial degradation of organic matter on the seabed is likely to reduce oxygen levels in bottom water under these circumstances. Dangerously low oxygen concentrations are more likely if the area is the site of a phytoplankton bloom, with dead phytoplankton raining down on the bottom, or if the area receives an input of organic wastes.

Oxygen depletion of bottom waters is a normal feature of parts of the Baltic and Caspian Seas (see Chapter 9) and summer thermoclines often develop in the central and eastern North Sea and in the New York Bight.

These are not usually damaging, but sometimes lead to widespread mortality of fish and the benthic fauna.

In spring and summer 1976, a strong thermocline developed over a large part of the middle Atlantic Bight (Fig. 2.14) accompanied by an intense phytoplankton bloom. Bottom oxygen concentrations were reduced over an area of 1200 km^2. It is estimated that 143 000 t of the clam *Spisula* were killed.

In summer 1981, a similar phenomenon was detected in a wide area of the German Bight and eastern North Sea (Fig. 2.15). In the worst affected areas almost no live fish were caught, most having left, but numerous dead flatfish (*Pleuronectes* and *Limanda*) were

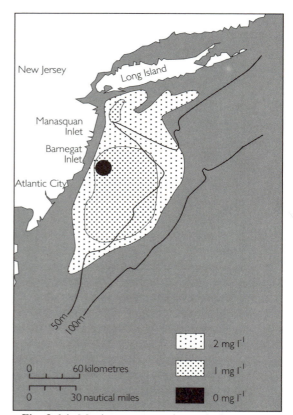

Fig. 2.14 Maximum extension of the area of reduced dissolved oxygen in bottom water of the Middle Atlantic Bight in summer 1976. The sludge dumping ground is in the northwest corner of the area. (*US Government Printing Office, Washington, DC.*)

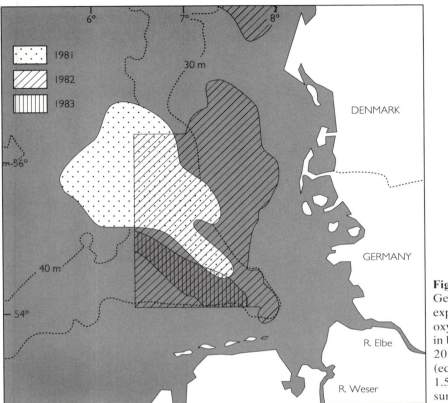

Fig. 2.15 Areas of the German Bight that experienced dissolved oxygen concentrations in bottom waters below 20 per cent saturation (equivalent to about 1.5 mg l^{-1}) during the summers of 1981–3.

observed on the seabed by underwater television. Brittle stars *Ophiura albida*, clams *Venus striatula*, and no doubt other burrowing fauna, were also killed. Simliar events were observed in this area in 1983 and 1986. The sewage sludge dumping grounds of the city of Hamburg were in this area, but this practice ceased in 1980 and the main source of plant nutrients in the area is the river inputs from the Elbe and Rhine.

River inputs and coastal discharges are probably responsible for the increasing eutrophication of the upper Adriatic (Fig. 2.16). Records of oxygen concentrations in surface and bottom waters since 1911 show that episodes of oxygen depletion in bottom waters first appeared in the mid-1950s off the delta of the river Po and that this area has extended eastwards in subsequent years. Increased oxygen saturation of surface waters

in the same areas is associated with high rates of primary production.

PUBLIC HEALTH RISKS

All human sewage contains enteric bacteria, pathogens and viruses, and the eggs of intestinal parasites. Contamination of food or drinking water may therefore pose a public health hazard. Generally, faecal contamination is measured by a count of the bacterium *Escherichia coli* (**coliform count**). *E. coli* is not a pathogen but is always present in the human intestine and appears in the faeces. The coliform count does not reflect the level of contamination by pathogens directly, merely the level of faecal contamination, but that in itself is a good measure of the risk to which the human population is exposed, depending on the general incidence of

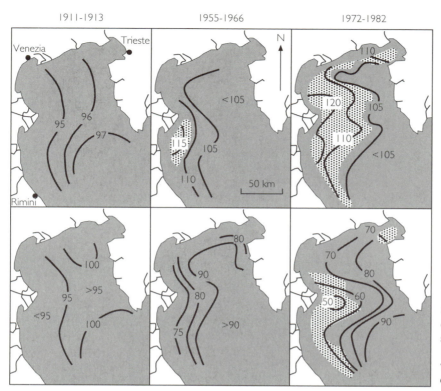

Fig. 2.16 Oxygen saturation (per cent) of surface waters (*upper*) and bottom waters (*lower*) in the upper Adriatic during the periods 1911–13, 1955–66, and 1972–82. Shaded areas experienced oxygen saturation above 120 per cent in surface water and below 20 per cent in bottom water.

diseases which can be communicated in this way.

Parasite eggs, including those of the nematodes *Ascaris* and *Ancylostoma* (hookworm) and of tapeworms such as *Taenia*, are resistant to drying and may persist in both crude sewage and sewage sludge. The use of either for agricultural use therefore presents some health risk, though more particularly for salad crops because the eggs are destroyed in the course of cooking vegetables. The problem is greatest in tropical countries and where the incidence of intestinal parasites in the human population is high.

Since humans do not drink seawater, it is generally assumed that there is no danger to sea bathers from faecal contamination of coastal waters unless the pollution is so gross that it is visible, in which case conditions would be so offensive that no-one would bathe there. ('If you cannot see it, it is safe.') This is an optimistic view; pathogenic bac-

teria are not necessarily killed by exposure to seawater, as was formerly supposed, but may enter a dormant phase so that they cannot be detected by normal methods (**non-platable bacteria**). They are, however, still capable of causing illness if ingested. Furthermore, sea bathing certainly occurs in some extraordinary places close to sewer outfalls, swimmers often swallow some seawater, and there have been some instances of hepatic or enteric disease being contracted through bathing in contaminated waters. EC regulations have set bacteriological standards for bathing beaches and a serious effort is being made in most European countries to reach these standards by introducing sewage treatment in coastal resorts or by extending submarine outfalls so that the discharge does not return to the beaches.

The chief health risk from sewage discharges to sea is undoubtedly through the ingestion of contaminated seafood. Filter-feeders, most conspicuously bivalve molluscs, flourish in waters enriched by a neighbouring input of organic wastes, but they accumulate the human pathogens on their gills and these may be transmitted to the consumer. The risk is obviously greatest if the bivalves are freshly harvested and eaten uncooked. Most bivalves in commerce which are collected from waters known to be contaminated by sewage effluents are, in fact, usually transferred to uncontaminated water for a few weeks before marketing (**depurated**), and during this time they lose their burden of pathogens and become safe to eat. Seafood organisms that are not filter-feeders, such as most crustaceans and fish, do not accumulate pathogens from sewage-contaminated water and do not represent a health risk from this source.

In its 1990 survey of the health of the world's oceans, the United Nations Group of Experts on the Scientific Aspects of Marine Pollution (GESAMP) placed sewage discharges to sea at the top of its list of concerns; heavy metal and oil pollution were given much lower priority. This was not because of the damage sewage causes in the marine environment, but because of the public health risk. In Third World countries with poor standards of public health and low nutritional status, enteric disease transmitted through seawater is responsible for many deaths, particularly among children.

Paralytic shellfish poisoning

Several species of the dinoflagellates *Gonyaulax*, *Gymnodinium*, and, in the Indo-Pacific region, *Pyrodinium* contain dangerous neurotoxins. In phytoplankton blooms ('red tides') composed of these species, high concentrations of the toxins can be accumulated by bivalves and some plankton-eating fish, generally without coming to harm themselves, but these then present a serious hazard to those that eat them. Fatalities have been recorded in sea birds and fish at the time of these blooms. People who eat contaminated fish or bivalves are at risk of suffering **paralytic shellfish poisoning** from the toxin with the symptoms of nausea, loss of balance, defective vision, and, in severe cases, convulsions and death. Areas regularly affected by toxic dinoflagellate blooms include the coasts of California and Florida, the Bay of Fundy, and the northeast coast of England, and shellfisheries have to be closed there during the summer, when blooms are most likely to occur. While adequate nutrients are necessary to support these toxic blooms, their location does not appear to be particularly related to high levels of organic inputs.

3

OIL POLLUTION

Oil pollution of the sea attracts great public attention because it is visible and most people encounter it, either at first hand on bathing beaches, or from pictures on television and in the press whenever there is a spectacular oil spill. Petroleum hydrocarbons reach the sea by many routes, however, and tanker accidents are by no means the only source of oil pollution.

INPUTS

It is difficult to calculate the total quantity of petroleum hydrocarbons entering the sea. Recent estimates vary between 1.7 and 8.8 m t yr^{-1}. The best estimate is about 2.5 m t yr^{-1} and the contributory sources are shown in Table 3.1.

Tanker operations

The world production of crude oil is about 3 b t yr^{-1} and half of it is transported by sea (Fig. 3.1). After a tanker has unloaded its cargo of oil it has to take on seawater as ballast for the return journey to the oil fields. The ballast water is usually stored in the cargo compartments which previously contained the oil. During unloading of the cargo, a certain amount of oil remains clinging to the walls of the compartments; this may amount to as much as 800 t in a 200 000 t tanker. The ballast water inevitably becomes contaminated with this 'clingage' which in any case has to be cleaned from all the compartments before they can be filled with a fresh cargo of

oil. Formerly this dirty ballast and clingage were discharged to sea during the return journey and were responsible for much of the casual oil pollution of the world's oceans. Two techniques have substantially reduced this: 'load-on-top' and 'crude oil washing'.

In the **load-on-top** system (Fig. 3.2), cargo compartments are cleaned by high pressure jets of water. The oily water is retained in the compartment until the oil floats to the top; the water underneath, containing only a little oil, is then discharged to the sea and the oil is transferred to a slop tank. This process is continued until all the compartments have been cleaned and the tanker carries only clean ballast water. At the loading terminal, fresh cargo is loaded on top of the oil in the slop tank, hence the name of the technique.

The load-on-top method reduces, but does not totally eliminate, the discharge of oil into the sea. Increasingly, the clingage is removed by jets of crude oil (**crude oil washing**) while the cargo is being unloaded. To avoid the danger of explosion of the petroleum vapour generated during this process, the empty compartments have to be flooded with inert gas derived from the engine exhaust. Some modern tankers have **segregated ballast**, that is the ballast water does not come in contact with oil, and there is pressure on tanker companies to ensure that all new tankers are constructed in this way.

With the introduction of these improved methods of deballasting, the amount of oil entering the sea as a result of tanker operations has steadily reduced from an estimated

Table 3.1 Estimated world input of petroleum hydrocarbons to the sea (m t yr^{-1})

Source		Total
Transportation		
Tanker operations	0.158	
Tanker accidents	0.121	
Bilge and fuel oil	0.252	
Dry docking	0.004	
Non-tanker accidents	0.020	
		0.555
Fixed installations		
Coastal refineries	0.10	
Offshore production	0.05	
Marine terminals	0.03	
		0.180
Other sources		
Municipal wastes	0.70	
Industrial waste	0.20	
Urban run-off	0.12	
River run-off	0.04	
Atmospheric fall-out	0.30	
Ocean dumping	0.02	
		1.380
Natural inputs		0.250
Total		2.365
Biosynthesis of hydrocarbons		
Production by marine phytoplankton		26 000
Atmospheric fall-out		100–4000

1 m t yr^{-1} or more in the mid 1970s, to 700 000 t in 1981, and to 158 000 t in 1989.

Dry docking

All ships, including oil tankers, require periodic dry docking for servicing, repairs, cleaning the hull, and so on. It is essential that all oil is removed from the cargo compartments of tankers and from the empty fuel tanks of all shipping, to avoid the risk of explosion from petroleum gases. Generally, the journey from the unloading terminal to the ship repair yard is a relatively short one and the load-on-top system cannot be used effectively. Slop reception facilities have to be provided at the ship repair yards. Formerly, many did not do so, but the situation has now

improved and the amount of oil reaching the sea in association with dry docking has been reduced from 30 000 t in 1981 to 4000 t in 1989.

Marine terminals

Accidents through human error and pipeline failure are an inevitable accompaniment to loading oil on to tankers and discharging it at oil terminals. The small size of this input reflects the care taken to reduce such accidents to a minimum.

Bilge and fuel oils

All shipping may need to take on ballast water when travelling unladen or in bad weather. Ballast tanks take up valuable cargo space and are limited in size, so additional ballast may be carried in empty fuel tanks. When the ballast water is pumped overboard it carries oil into the sea. In addition, all shipping needs to pump out bilge water which invariably contains oil from the ship's engines. This oil can be removed by separators, which is required in some sea areas under international conventions, though there are undoubtedly many illegal discharges of oily bilge water. Individually, the quantity of oil released may be small, but since all shipping contributes, the total amount of oil entering the sea is considerable.

Tanker accidents

As with other shipping, a large number of accidents involving oil tankers happen every year. Most result in no, or quite trivial, spillages of oil; the accident may not rupture the cargo compartments, a damaged vessel may be salvaged, or its cargo may be off-loaded into other tankers. Major shipping disasters involving oil tankers are a different matter. In March 1978, the *Amoco Cadiz*, carrying 223 000 t of Arabian and Kuwaiti crude oil, was wrecked on the coast of Brittany and lost its entire cargo, causing immense damage over a wide area. Fortunately, such catastrophes are rare events.

Because shipping hazards are greatest close to shore and in narrow straits, such as the

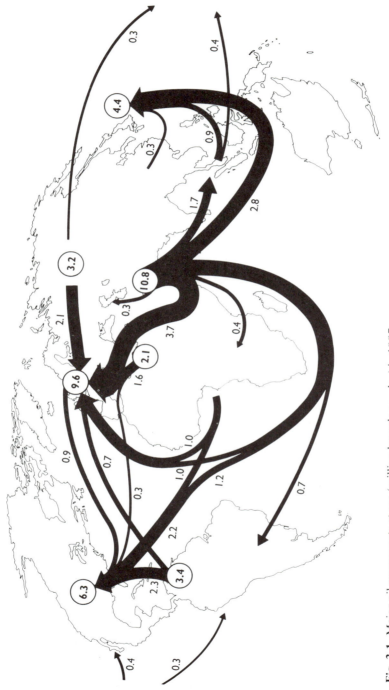

Fig. 3.1 Major oil movements at sea (million barrels per day) in 1987.

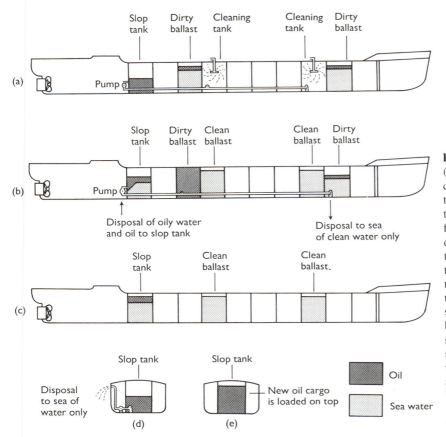

Fig. 3.2 Load on top: (*a*) empty tanks are cleaned by water jet and the washings transferred to the slop tanks, (*b*) oil floats to the top in tanks containing dirty ballast; the water is discharged to the sea and oil transferred to the slop tank, (*c*) eventually the ship carries only clean ballast, and oil in the slop tanks floats to the surface, (*d*) underlying water is pumped out, and (*e*) new cargo is loaded on top of the oil remaining in the slop tank.

Straits of Dover, or near the entrances to ports where the density of shipping is high, most tanker accidents occur close to the shore and if oil is spilled, coastal pollution almost invariably results.

Non-tanker accidents

When a ship is in an accident, its fuel oil may be lost to the sea. Some cargo ships, particularly bulk carriers, are now very large and carry as much fuel oil as a 1960 oil tanker carried crude oil, so this source of oil contamination is not negligible.

Offshore oil production

The oil that is extracted from the seabed invariably contains some water (**production water**) which must be extracted before the oil is transported to the refinery. This is done by oil separators on the platform and the oil

concentration in the water that is discharged is usually less than 40 parts per million (p.p.m.), but in aggregate this amounts to a considerable quantity.

When an oil well is being drilled, **drilling muds** are pumped down the well. These maintain a head of pressure and prevent a blow-out when oil is struck, lubricate the drill bit, and carry the cuttings back to the surface (Fig. 3.3). The drill muds contain water or, if the nature of the strata demands it, 70–80 per cent oil; formerly the oil was diesel, but increasingly low-toxicity oils are used. While there is some attempt to separate the drill cuttings from the **oil-based muds** before they are dumped on the seabed beneath the platform, they are inevitably still heavily contaminated and 90 per cent of petroleum hydrocarbons entering the North Sea as a result of offshore oil extraction is from this

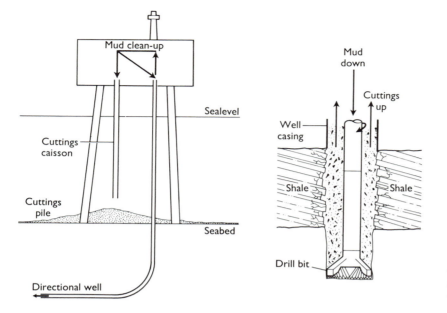

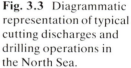

Fig. 3.3 Diagrammatic representation of typical cutting discharges and drilling operations in the North Sea.

source. There are now moves to ensure that oil-contaminated cuttings are brought ashore for disposal.

Blow-outs, the uncontrolled release of oil from the well, are catastrophes and great precautions are taken to prevent them. Nevertheless, accidents happen occasionally and great quantities of oil may be released to the sea before the blow-out is brought under control. In April 1977, a well in the Ekofisk field in the Norwegian sector of the North Sea blew out and discharged 20 000–30 000 t of oil. In June 1979, a well in the Ixtoc field off the Mexican coast blew out and was on fire. It took nine months to bring it under control and in that time it is estimated to have discharged 350 000 t of crude oil into the Gulf of Mexico. Gas blow-outs, as in the Piper field in the North Sea in 1988, are less likely to result in oil pollution.

Atmosphere

The incomplete combustion of petrol or diesel in motor vehicles results in petroleum hydrocarbons being released into the atmosphere. They are washed out in rain either directly into the sea or indirectly by contributing to river run-off.

Municipal and industrial wastes

Domestic wastes and sewage contains a quantity of oils and greases and, depending on the nature of the industry, its wastes may also contain a considerable quantity of petroleum hydrocarbons. In coastal areas, these wastes are often discharged into the sea and, even if subjected to treatment, the sewage sludges may retain petroleum hydrocarbons which are then discharged to sea.

Coastal oil refineries

Older refineries use a steam-cracking process (see p. 44) and the water recovered, which is discharged in the effluent, contains up to 100 p.p.m. of oil in the water. In more modern refineries, the oil does not come in contact with the water and it is possible to reduce the quantity of oil in the effluent to 25 p.p.m. Refineries require a large volume of water and the total discharge of oil is not negligible, especially as it is continuously discharged into the same body of water.

Urban and river run-off

Every time it rains, irridescence caused by oil and petrol can be seen on the roads. This is washed down drains and into water courses and eventually reaches the sea. Garage forecourts sustain a large amount of spilled oil which is washed into the drains.

Success in recovering used engine oil purchased for use in motor vehicles varies widely. In Germany and the Netherlands, virtually all such oil is recovered and refined or used as a fuel. In Britain, France, and Italy only 25–30 per cent of that sold is recovered. With motorists increasingly making their own oil changes, it must be suspected that much of this 'lost' oil finds its way into drains or on to the land, from which it is carried into rivers and then to the sea.

Licensed dumping at sea

Shipping channels in estuaries and ports commonly need regular dredging. The dredging spoil, which is usually dumped at sea, is contaminated with oil. Various kinds of solid municipal and industrial wastes that are dumped at sea may also contain petroleum hydrocarbons.

Natural sources

Oil deposits close to the earth's surface seep out and have done so for millennia. The pitch lakes of Trinidad are the product of natural seepage, and coastal oil seeps occur in many parts of the world.

Biosynthesis

Oil deposits are produced by plant remains that have become fossilized under marine conditions (coal results under freshwater conditions). Living plants produce hydrocarbons; the aromatic scent of pinewoods is caused by them. Estimates of the annual production of hydrocarbons by marine phytoplankton and by land plants from which the products are rained out into the sea are inevitably vague but, in any case, the figures dwarf the input of fossil hydrocarbons by several orders of magnitude. Recent and fossil hydrocarbons have different constitutions and may well have different effects on marine ecosystems, but the dominating input of hydrocarbons from plants needs to be borne in mind when assessing the effect of petroleum hydrocarbons.

WHAT IS OIL?

Crude oil is a complex mixture of hydrocarbons with 4–26 or more carbon atoms in the molecule (Fig. 3.4). Arrangements include straight chains, branched chains, or cyclic chains including aromatic compounds (with benzene rings). Some polycyclic aromatic hydrocarbons (PAH) are known to be potent carcinogens. Sulphur and vanadium compounds are also included in crude oil and non-hydrocarbons may represent up to 25 per cent of the oil. The exact composition of crude oil varies from one oilfield to another: much of the North Sea oil is light, with little sulphur and low in tars and waxes; oil from the Beatrice field in the Moray Firth and that to the west of Shetland is a very heavy waxy oil which needs to be heated in order to pump it through pipelines. The composition of the crude oil also varies during the life of a single oilfield.

Crude oil must be refined before it can be used. Refining is essentially a distillation process with different **fractions** or **cuts** taken at different boiling ranges (Table 3.2). Light gasoline is the basis for petrol used in motor vehicles; naphtha provides feedstock for the petrochemical industry; the residue is used as bunker fuel in ships and power stations; and the higher fractions are used as tars, and so on. Many of the commercial products are further refined, made into particular formulations, and receive additives of other materials to suit them for their various purposes.

All components of crude oil are degradable by bacteria, though at varying rates, and a variety of yeasts and fungi can also metabolize petroleum hydrocarbons. Small, straight- and branched-chain compounds degrade most rapidly, cyclic compounds the slowest. High-molecular-weight compounds, the tars, degrade extremely slowly.

(a) Methane (CH$_4$) the
simplest hydrocarbon

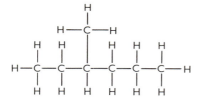

(b) A straight-chain alkane (or
paraffin):heptane (C$_7$H$_{16}$)

(c) A branched-chain alkane

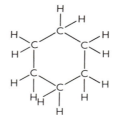

(d) A cyclo-alkane
(naphthene)

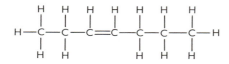

(e) An unsaturated hydrocarbon

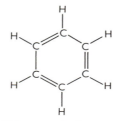

(f) Benzene, the simplest
aromatic hydrocarbon

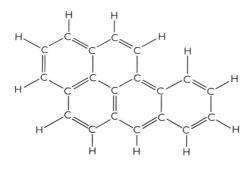

(g) Benzo [a] pyrene, a
polycyclic aromatic
hydrocarbon (PAH)

Fig. 3.4 The structure
of some hydrocarbons.

Table 3.2 Refinery 'cuts' of crude oil

	Boiling range (°C)	Molecular size (Number of carbon atoms)
Petroleum gases	30	3–4
Light gasoline, benzine	30–140	4–6
Naphtha	120–175	7–10
Kerosene	165–200	10–14
Gas oil (diesel)	175–365	15–20
Fuel oil and residues	350	20+

FATE OF SPILLED OIL

When liquid oil is spilled on the sea, it spreads over the surface of the water to form a thin film—an **oil slick**. The rate of spreading and the thickness of the film depend on the sea temperature and the nature of the oil; a light oil spreads faster and to a thinner film than a heavy waxy oil.

The composition of the oil changes from the time it is spilled (Fig. 3.5). Light (low-molecular-weight) fractions evaporate, water-soluble components dissolve in the water column, and immiscible components become emulsified and dispersed as small droplets. The rate of emulsification of the oil in water depends on the agitation provided by waves and water turbulence. In some sea conditions, a water-in-oil emulsion is produced; this may contain 70–80 per cent of water and forms a viscid mass, known from its appearance as **chocolate mousse**. This forms thick pancakes on the water and intractable sticky masses if it comes ashore. The heavy residues of crude oil form **tar balls**, ranging in size from less than 1 mm to 10–20 cm in diameter.

Emulsified oil is in microscopic drops and therefore presents a large surface area at which bacterial attack can take place. Tar balls and mousse present a small surface area compared to their volume, and degrade extremely slowly for this reason. Tar balls of various sizes occur in the surface waters of all oceans, though they predominate in major shipping routes and in major ocean currents such as the Gulf Stream. They are mostly derived from tanker washings and routine shipping operations.

An oil slick does not remain in one place but travels downwind at 3–4 per cent of the wind speed (Fig. 3.6) except in enclosed waters and estuaries where tides and water currents have a greater influence on movements of the slick. When a slick encounters land it is stranded on the shore with well-known consequences.

TREATMENT OF OIL AT SEA

Cleaning oil from beaches is difficult, time-consuming, labour-intensive, and costly. There are obvious advantages if an oil slick that threatens the coast can be removed or otherwise dealt with while it is still at sea. A variety of techniques to do this has been developed, but only three have proved effective and even these can rarely prevent beach pollution if a large amount of oil is spilled close to shore.

The natural process of emulsification of oil in the water can be speeded up by spraying chemical **dispersants** on the oil slick from ships or aircraft. Dispersants are not effective against heavy oils or oil that has been on the sea for some time ('weathered oil'). The first dispersants to come into use were highly toxic and there was some reluctance to use them,

Time (hours)

0   1   10   day   100 week month 10^3 year 10^4

Spreading

Drift

Evaporation C_6 C_7 C_8 C_9 (aromatics)

                                      C_{14} C_{15} (n - alkanes)

                        C_{11} C_{12} C_{13}

Dissolution   Vertical   Horizontal

Dispersion

Emulsification

Sedimentation

Biodegredation

Photooxidation

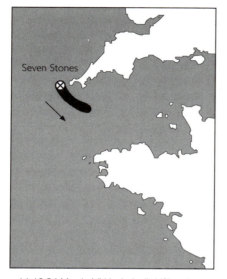

(a) 18-24 March, NW wind, oil drifting at sea

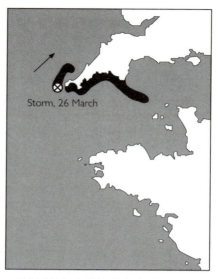

(b) 24-26 March, SW wind, oil ashore south Cornwall

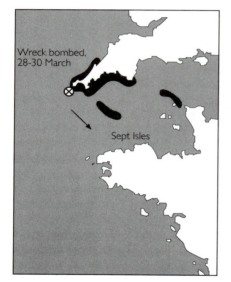

(c) 26 March-7 April, NW wind, oil ashore north Cornwall

(d) 8-12 April, NE wind, oil ashore north Brittany

Fig. 3.6 Movement of oil from the wrecked tanker *Torrey Canyon* in response to changing wind direction.

but low-toxicity dispersants have now replaced them. While dispersants are useful for treating small quantities of fresh oil, in any large spill their chief limitation is the difficulty of deploying sufficient ships or aircraft to spray more than a small proportion of the oil.

Seawater intakes, mariculture installations, or other important sites can be protected by the use of floating **booms**, with a 'sail' above the water-line and a 'skirt' below (Fig. 3.7), to deflect the floating oil to less sensitive areas. Large V-shaped booms may also be used to corral the oil, which can then be pumped out where it accumulates at the point of the V.

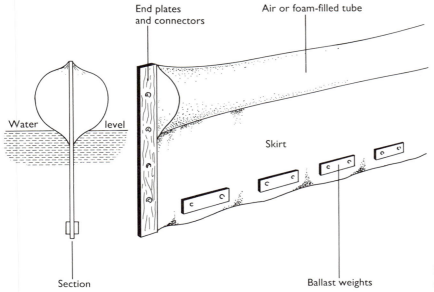

End plates
and connectors

Air or foam-filled tube

Water   level

Skirt

Section

Ballast weights

Fig. 3.7 Design of one
type of floating boom.

A variety of **slick-lickers** (Fig. 3.8) has been developed in which a continuous belt of absorbent material dips through the oil slick and is passed through rollers to extract the oil. Slick-lickers can deal with only small quantities of oil and are useful in harbours and sheltered waters rather than the open sea.

BEACH CLEANING

If a large quantity of oil is stranded and the shore is accessible, much of the oil can be pumped into road tankers and removed, thus reducing the chance of the oil being removed on a succeeding tide and redeposited elsewhere, thereby extending the area that is contaminated. Even when it is possible to recover some of the oil from the beach, however, a considerable amount remains on and between rocks and drained into the substratum. The techniques to remove this depend on the nature of the shore and since the treatment is often more damaging to the fauna and flora than the oil itself, it is often preferable not to attempt to clean up an oiled beach. The decision to clean or not to clean depends on the value placed on the beach. A tourist amenity beach has a high priority for cleaning because the fauna and flora is of far less interest than the preservation of oil-free conditions.

Rocks, harbour walls, and similar surfaces may be cleaned by high-pressure hoses, high-pressure steam, or dispersants. If chemical agents are used, they must be accompanied by a large volume of water in which the oil can be dispersed. To achieve this, rocks are sprayed with dispersant ahead of a rising tide, when breaking waves provide the agitation and water volume to remove the oil. Alternatively, the rocks must be hosed down after the dispersant has been applied.

Dispersants are useless on pebble or sandy beaches because the dispersed oil merely drains into the beach; it disappears from the surface only to reappear at some later date. The only means of cleaning such beaches is to remove the oiled surface layers of the substratum, either manually or by bulldozer.

These are drastic treatments and the damaging effects on the fauna and flora can be reduced by using straw or cut vegetation as an absorbent to mop up much of the oil. On sheltered rocky shores with a good algal growth, a large amount of oil is trapped in the seaweeds, which can be cut and gathered. Physical removal of the oil by these methods

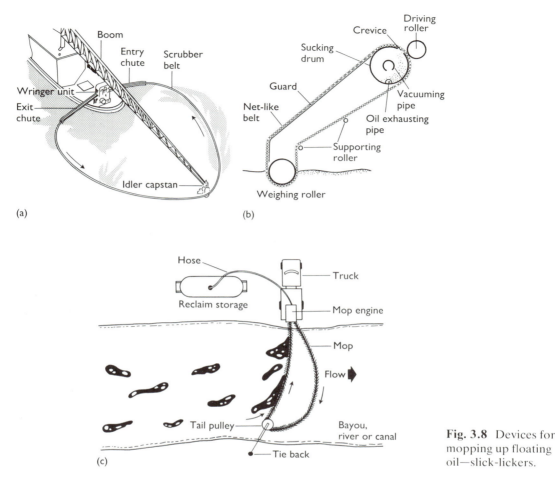

(a)

(b)

(c)

Fig. 3.8 Devices for mopping up floating oil—slick-lickers.

results in only partial cleaning and much oil remains.

Unless beach oil can be dispersed into the sea, any beach cleaning operation produces a large volume of oil-contaminated debris. A few tonnes of oil can easily result in hundreds of tonnes of oily sand, pebbles, and other debris, and disposal of this material presents a serious problem. It cannot be incinerated unless the oil content is very high; it is usually unsuitable for dumping on waste tips because of the risk of the oil leaching out and contaminating water courses; and oil refineries cannot deal with it. It is possible to plough oily material into land where bacterial degradation eventually disposes of it, but this is unlikely to be an effective way of dealing with large quantities of oiled material. At

present, tipping on waste dumps where the oil can be contained appears to be the only practicable, if unsatisfactory, solution.

TOXICITY OF PETROLEUM HYDROCARBONS

Water-soluble components of crude oils and refined products include a variety of compounds that are toxic to a wide spectrum of marine plants and animals. Aromatic compounds are more toxic than aliphatics, and middle-molecular-weight constituents are more toxic than high-molecular-weight tars. Low-molecular-weight compounds are generally unimportant because they are volatile and rapidly lost to the atmosphere. A spillage of diesel fuel, with a high aromatic content, is therefore much more damaging than bunker fuel and weathered oil, which have a low aromatic content. A spillage of petrol or other 'white spirit' may present a serious fire hazard, but has little impact on marine organisms in the water.

ECOLOGICAL IMPACT OF OIL POLLUTION

Rocky substrata

Spectacular changes on some rocky beaches in southwest England, following the wreck of the tanker *Torrey Canyon* off Land's End in March 1967, drew public attention to the great impact a very large oil spill could have. It caused concern about the effect of pollution in general on the health of the oceans, and it stimulated a large research effort in marine pollution in the following years.

Rocky shores are high-energy beaches and stranded oil is quickly removed from the intertidal region by wave action and water movement. This is not necessarily an advantage if fresh oil is then redeposited on other beaches, repeating the damage elsewhere. Oil is removed most slowly from extreme high- and low-water levels and sheltered crannies, where wave energy is least. Depending on the nature and age of the oil, a considerable variety of animals and the more sensitive red and green algae are killed; but much of the oil reaching the beaches is crude oil or bunker fuel which has been at sea for some days and lost most of its toxic constituents, and it poisons few organisms on the beach. Limpets, *Patella*, have even been observed to graze dried oil from the rocks without coming to harm. But even if it is relatively non-toxic, weathered oil may cause damage through its physical properties. Many seaweeds secrete mucins which prevent oil adhering to them, but if it does adhere to the fronds, seaweeds can then be torn from their stipes during storms because of their increased weight. Large amounts of stranded oil may kill animals by smothering them. On shores that received oil from the Santa Barbara blow-out in California, large *Balanus perforatus*

projected above the oil and survived, but smaller species of barnacles were killed by smothering.

The conspicuous damage on some Cornish beaches following the wreck of the *Torrey Canyon* was, ironically, not due to the stranded oil but to the very toxic oil-spill dispersants then in use which had, in the confusion, been poured neat over a few beaches. The dispersant or the mixture of oil and dispersant proved far more lethal than the oil alone and most organisms on the affected beaches were killed. Most critically, the limpets *Patella* were eliminated. In the absence of these herbivores, diatoms and algae colonized the rocks and an algal succession continued unchecked during the summer, culminating in a dense stand of *Fucus*. The presence of *Fucus* inhibited a settlement of *Patella* or barnacles the following season and the change from a community dominated by *Balanus* and *Patella* to one dominated by algae appeared to be long standing. Some four to five years later, *Patella* became re-established and, surprisingly, were able to graze down the *Fucus* by rasping at the stipe, which became weakened and the fronds were then torn off. More than ten years after the spill, although barnacle domination of the rocky shore community had been restored, the associated fauna still appeared to be less rich and varied than before.

A similar consequence of the loss of dominant herbivores has been recorded in a sub-littoral rocky community. In March 1957, a small tanker, the *Tampico Maru*, was wrecked and lost its cargo of diesel fuel at the mouth of a small bay on the Mexican Pacific coast of Baja California. The dominant sub-littoral herbivores were the abalone *Haliotis* and two species of the sea urchin *Strongylocentrotus*, and these were either killed or left the affected area. A dense growth of giant kelp *Macrocystis* followed and persisted in the bay for five years before the herbivores were re-established and brought it under control (Fig. 3.9).

In both the Cornish and Mexican examples, restoration of the original community was

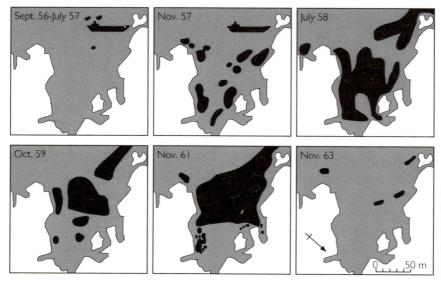

Fig. 3.9 Growth of the giant kelp *Macrocystis* (shown in black) following pollution by diesel fuel from a wrecked tanker on the Mexican Pacific coast.

delayed because of the elimination of the dominant herbivores, the establishment of an algal cover during the period when there were no colonizing larvae of the herbivores, and then the presence of a dense stand of algae inhibiting recolonization by herbivores in the following season. In both cases, such algal domination persisted for several years and, at least on the Cornish beaches, evidence of this disturbance could still be detected a decade after the original oil spill. Similar consequences of the loss of herbivores, and a similar time-course of recovery, have been confirmed on many occasions subsequently. Less critical damage, or damage at such a time that the herbivores can be re-established before there has been too dense a growth of algae, is made good much more rapidly. Even Cornish beaches affected by oil from the *Torrey Canyon* were, in many cases, in an essentially normal condition again within two years.

Soft substrata

Stranded oil is not readily removed from low-energy sedimentary beaches and, if still liquid, it drains down into the substratum. Here the low oxygen concentration does not favour bacterial degradation of the oil, which may therefore retain its toxic properties for some time. When the *Amoco Cadiz* was wrecked on the Brittany coast in March 1978, releasing 223 000 t of crude oil, a considerable amount of it was carried into estuaries and inlets where it became incorporated in sediments. In addition to the immediate damage to the fauna, the persistence of toxicity prevented the start of recovery processes, and oil leaching from the sediments a year later caused renewed contamination.

In September 1969, the barge *Florida* ran aground during a storm in Buzzard's Bay, Massachusetts, near West Falmouth and spilled 10 000 gallons (nearly 40 000 l) of diesel fuel. Although this was a relatively small oil spill, the toxic nature of the oil (containing 41 per cent aromatics) and the circumstances of the spillage made it unusually damaging. Winds and strong surf drove the oil into Wild Harbor, on to beaches, and churned up bottom sediments with which the oil became incorporated and was carried down to the bottom. Subsequent sediment transport extended the contaminated area. There was an immediate kill of fish particularly in shallow creeks and bays which shelter juveniles of commercial species such as flounders and blue fish, and the adults of a variety of bait fish. Lobsters, crabs, shrimps, and bivalves were killed in large numbers.

Scallops were particularly badly affected; oysters and soft-shell clams less so. Commercial shellfish beds had to be closed because of tainting on a long-term basis due to the continued risk of contamination by oil released from the shifting sediments. Detailed studies of the sub-tidal benthic community revealed instability persisting for more than five years in the most polluted parts of Wild Harbor; less contaminated areas began to show a successional recovery two years after the spillage.

Plankton

Plankton, and especially the neuston living in the top few centimetres of the sea, might be supposed to be particularly at risk because it is exposed to the highest concentration of water-soluble constituents leaching from floating oil.

Oil and oil fractions are toxic to a wide range of planktonic organisms, aromatic compounds more so than aliphatic ones. Weathered oil, after loss of volatile and water-soluble components, is not very toxic and, indeed, copepods have been observed to ingest oil particles and pass them through the gut without harm. Very low concentrations of petroleum hydrocarbons, below 50 ng g^{-1}, enhance photosynthesis, presumably because they have a nutritive effect. Above 50 ng g^{-1} there is a progressive depression of photosynthesis in algal cultures. At concentrations above 250 ng g^{-1}, feeding in the copepod *Acartia* is depressed and food selection is altered.

Studies of the response of whole plankton communities in the natural environment following oil spills yield conflicting results, but do not support the gloomy predictions from the laboratory studies. Following an oil spill from the tanker *Tsesis* in Swedish waters in 1977, zooplankton biomass fell dramatically for about five days by the death of the animals or their avoidance of the area. In the absence of the grazing zooplankton, there was a marked increase in phytoplankton biomass and productivity; it is not clear whether the oil had any nutritional effect. Following the wreck of the *Amoco Cadiz* in 1978, and the massive oil pollution of the Brittany coast, whatever initial impact there may have been on plankton, there was no sign of abnormality two to three weeks later, although there is a suggestion that the spring plankton outburst was depressed for a period.

Although it cannot be doubted that oil pollution, if sufficiently severe, kills planktonic organisms, it has not been possible to detect more than very transient effects and sometimes not even those. Whatever losses are incurred are evidently rapidly made good, either by renewed growth or immigration from unaffected areas. Since petroleum hydrocarbons are subject to degradation by yeasts and bacteria, their main effect is to enhance primary production and act as a nutritive source.

Fixed vegetation

Salt marshes and, in the tropics, mangrove swamps are, like intertidal mud-banks, low-energy areas likely to trap oil, and the plants which form the basis for these ecosystems suffer accordingly. Both are important ecosystems at the boundary between land and sea. They control coastal erosion, are a source of organic production which is transferred to the sea, and they provide shelter for the young stages of marine organisms, some of commercial value.

The effect of oil pollution on annual plants living in a salt marsh depends on the season: if the plants are in bud, flowering is inhibited; if the flowers are oiled they rarely produce seeds; and if the seeds are oiled, germination is impaired. Generally it may be expected that annuals will be killed by oiling and be dependent on reseeding from outside the area; recovery of annual populations may therefore require two or three seasons. Perennials show a range of reactions. Shallow-rooted plants with no or small food reserves, such as *Sueda maritima* or *Salicornia*, are readily killed. Others will generally survive at least a single exposure to oil and perennials with large food reserves, such as those with taproots, survive repeated oilings. Generally the foliage is cut back, but decomposing oil has a nutrient effect and there is rapid and luxuriant renewed growth (a 'flush'). Experience of isolated oil spills suggests that oil pollution of this kind is less damaging to salt marshes than efforts to clean up the oil.

Mangroves present a rather different problem. They live in anoxic muds and have extensive air spaces carrying oxygen to the submerged part of the tree. Lenticels, by which the air is taken up, occur on the aerial roots of *Avicennia* or prop roots of *Rhizophora* and, if the lenticels are clogged with oil, the oxygen level in the root air spaces falls to 1–2 per cent of normal within two days. Although mangroves have certainly suffered damage by oil spills, there are a number of cases where heavy oilings have not killed the plants. Too few studies of these tropical ecosystems have been made to allow a proper assessment to be made of their vulnerability to oil pollution.

Sea birds

Whatever other effects oil pollution may have, the loss of sea birds attracts the greatest public concern. It is difficult to give a precise estimate, but it is quite possible that tens or even hundreds of thousands of sea birds are oiled in the northeast Atlantic every year. This sort of toll has continued ever since shipping converted from coal to oil-fired boilers around the time of the First World War and motor cars became numerous in the 1920s. It is feared that sea birds might show a population decrease as a result of this heavy and persistent mortality.

Unlike most other organisms in the sea, sea birds are harmed through the physical properties of floating oil, and the toxicity of its constituents is of minor importance. If liquid oil (or any other surface-active substance) contaminates a bird's plumage, its water-repellent properties are lost. If the bird remains on the sea, water penetrates the plumage and displaces the air trapped between the feathers and the skin. This air layer provides buoyancy and thermal insulation. With its loss, the plumage becomes waterlogged and the birds may sink and

drown. Even if this does not happen, the loss of thermal insulation results in a rapid exhaustion of food reserves in an attempt to maintain body temperature, followed by hypothermia and, commonly, death. Birds attempt to free their plumage of contaminating oil by preening and they swallow quantities of it. Depending on its toxicity, the oil may then cause intestinal disorders and renal or liver failure. Quite small quantities of oil ingested by birds during the breeding season depress egg-laying and, of the eggs that are laid, the proportion that hatch successfully is reduced. If oil is transferred from the plumage of an incubating bird to the eggs, the embryos may be killed.

Indirect effects of oil pollution on reproduction appear to be much less important than the direct mortality of adult birds, and most attention has been directed towards the latter problem. The species most commonly affected are auks: guillemots (murre), razorbills, and puffins; and some diving sea ducks: scoters, velvet scoters, long-tailed ducks (old squaw), and eiders. These birds spend most of their time on the surface of the water and so are particularly likely to encounter floating oil, and because they dive rather than fly up when disturbed, they are as likely as not to resurface through the oil slick, so becoming completely coated with oil. Furthermore, these ducks are extremely gregarious except when ashore for breeding, and the auks are gregarious at all times of the year. Thus, if there are casualties they are likely to be numerous. Indeed, quite small oil slicks drifting through concentrations of birds resting on the sea may inflict very heavy casualties quite disproportionate to the quantity of oil. Thus, when 230 000 t of crude oil was lost from the *Amoco Cadiz* on the Brittany coast, the known seabird casualties numbered 4572. On the other hand, the much smaller oil spill of 35 000 t of crude oil following the grounding of the *Exxon Valdez* in Alaska in March 1989 resulted in over 30 000 known seabird deaths and probably very many more. Almost as large a death toll occurred in the Skagerrak in January 1981, when 30 000 oiled birds appeared on beaches. This appears to have been caused by very small amounts of oil discharged by two vessels.

Other species besides auks and sea ducks are also regularly affected, though usually in much smaller numbers because of their habits and because they do not occur in such dense flocks. These include grebes and divers (loons), shags and cormorants, mergansers, gannets, and pelicans. The flightless penguins in the southern hemisphere occupy much the same ecological niche as guillemots in the northern hemisphere but, except for the jackass penguin around the coast of South Africa, the centre of population of most penguins is well to the south of heavily trafficked sea lanes or other sources of oil pollution and so they rarely encounter oil slicks.

It is usually impossible to estimate more than crudely the number of birds killed by oil pollution. The only reliable figures are of the number of oiled birds or carcasses found on the shore, although an unknown proportion of the carcasses may have become oiled after death from other causes. An unknown number of oiled birds never reach the shore and, depending on wind speed and direction, sea conditions, distance from shore of the bird flocks, and accessibility of the coast to observers, it is likely that counts of oiled birds coming ashore underestimate actual casualties by anything up to 90 per cent. Figure 3.10 shows winter sea bird casualties around North Sea and English Channel coasts, and the values reflect the interaction of these factors. The high proportion of birds stranded on the coast of Holland that are found oiled, for example, reflects the prevalence of westerly winds in winter, drifting both the bird flocks and any floating oil to the eastern side of the North Sea. The exceptionally heavy loss of sea birds in the Skagerrak in January 1981 followed a period of westerlies which had funnelled large numbers of wintering birds into those narrow waters.

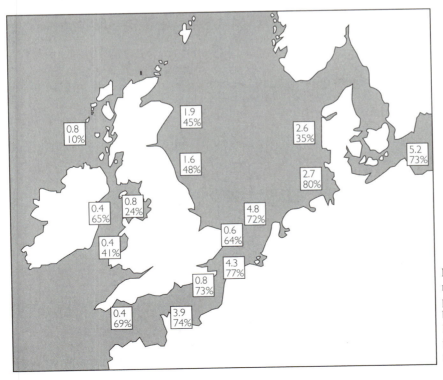

Fig. 3.10 Average number of dead birds per kilometre found in beach surveys in late February, 1967–73, and the percentage of them that had been oiled.

The auks not only include species which suffer most heavily from oil pollution, but they also have a reproductive pattern which does not favour rapid recovery from heavy adult losses. Guillemots, for example, do not start to breed until they are 3–4 years old and thereafter do not breed every year; they produce only a single egg and the risk of losing eggs or chicks from predators or simply from rolling off precarious cliff ledges is such that only about 0.2 chicks per year are sucessfully reared for each breeding bird. The decline in guillemot and puffin colonies in southwestern Britain, Brittany, and on the New England coast that had continued for much of this century, appeared to confirm that steady losses from oil pollution could not be made good by natural reproduction. Similar fears were expressed for the population of those species of sea duck particularly exposed to oil pollution, but since these produce numerous young there is less justification for concern.

Contrary to what had been predicted, while southerly colonies of puffins, guillemots, and razorbills were declining, more northern colonies of these and other sea birds have shown a dramatic increase in the last one or two decades, despite continuing oil pollution. The decline of southern colonies appears to have been the result of climatic changes in the north Atlantic in the period 1850–1950. The reasons for the growth of more northerly colonies is uncertain but appears to be a consequence of young birds starting to breed at an earlier age than before. It may be noted that occasional catastrophic mortality is not unusual for arctic and subarctic sea birds—a long period of severe weather which prevents feeding can cause as high a mortality as the worst oil spill—and despite their low reproductive potential, it would be surprising if a biological mechanism did not exist to maintain the population in the face of erratic losses of this kind.

Marine mammals

Although there have been occasional reports of seal pups being severely oiled and possibly killed by crude or bunker oil, there is no evidence that adult seals and sea lions or cetaceans appear particularly at risk from oil on the sea.

Sea otters (*Enhydra lutris*) are exceptional among sea mammals. Unlike seals and cetaceans, which rely on subcutaneous blubber to provide thermal insulation, sea otters rely on their dense fur which functions in a similar way to the plumage of a sea bird and they are similarly vulnerable to floating oil. Once driven to the verge of extinction by fur hunters, they are now common on the northeast Pacific coast, although rare elsewhere. A thousand or more were killed in the *Exxon Valdez* oilspill, in 1989, but with their high reproductive rate (populations are capable of growing at 17–20 per cent per annum until they reach a ceiling imposed by their food supply), the Alaskan populations are expected to recovery quickly.

IMPACT OF REFINERY WASTE WATER

Oil refineries discharge waste water containing some petroleum hydrocarbons. The receiving waters are therefore subject to low-level, chronic pollution.

The refinery at Fawley in Southampton Water used a steam-cracking process in which water comes in contact with oil and the effluent inevitably has a high oil content. From 1953, two effluent streams were discharged into creeks running across a salt marsh system. As well as oil and chemicals normally present in refinery effluent, oil from accidental spillages in the refinery and at the unloading jetty entered the stream from time to time. Between 1953 and 1970, these discharges caused the loss of vegetation from a considerable area of the marsh. In 1970 a programme of effluent improvement was instituted and, by 1975, the oil content of the effluent had been reduced from 31 p.p.m. to 10 p.p.m. and the quantity of the discharge from $28\,000$ m^3 h^{-1} to $21\,000$ m^3 h^{-1} (Table 3.3). The salt marsh was monitored from

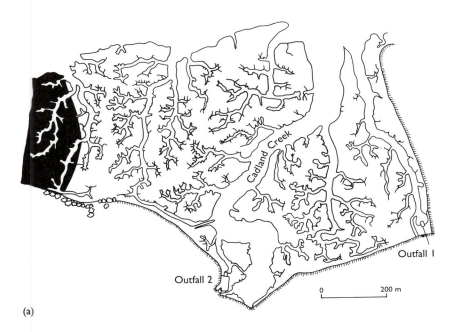

(a)

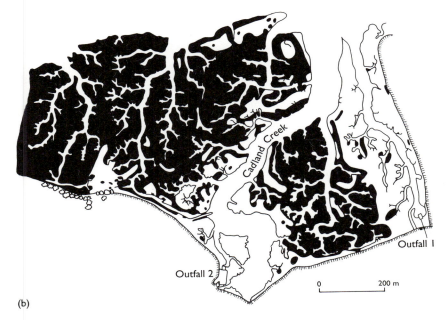

(b)

Fig. 3.11 Fawley salt marsh: (*a*) distribution of vegetation (black) in 1970, (*b*) distribution of *Salicornia* in 1980.

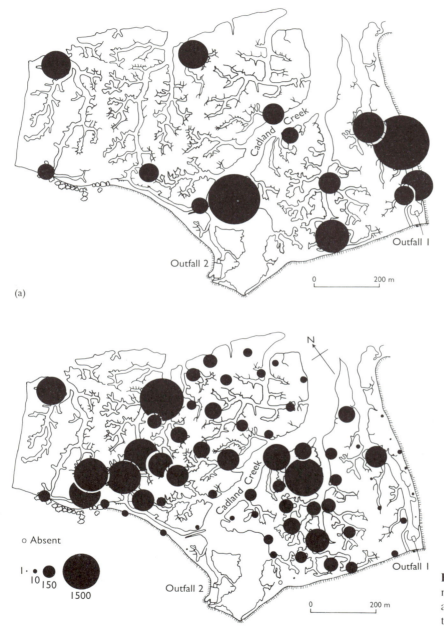

(a)

o Absent

1· · 10 150 1500

Fig. 3.12 Fawley salt marsh: (*a*) hydrocarbons and (*b*) oligochaetes in the substratum in 1980.

Table 3.3 Quality of effluent from Fawley refinery

	Average oil content (p.p.m.)	Discharge rate (m³ h⁻¹)
1963–70	31	28 000
1972	25	28 000
1974	14	23 000
1979	10	21 000

1970 onwards and by 1980 vegetation had become re-established over a considerable part of the area (Fig. 3.11). The sediments, however, remained contaminated and in 1980 their content of aliphatic hydrocarbons ranged from 340 to 17 750 p.p.m. The fauna of the area, previously denuded of vegetation, remains impoverished and consists mainly of enchytraeid and tubificid oligochaetes. They are found in large numbers in soil containing up to 4750 p.p.m. aliphatic hydrocarbons, but in small numbers in the heavily contaminated areas (Fig. 3.12). Inshore sediments are also contaminated and contain reduced numbers of the polychaete *Nereis diversicolor*.

Experience at Milford Haven, Pembrokeshire is in marked contrast to that at Fawley. It is the site of the largest deep-water oil terminal in Europe and five refineries have been constructed around it, all discharging waste water into the Haven. It has an exceptionally rich marine fauna and flora and is part of the Pembrokeshire Coast National Park; because of this, unusual care is taken to minimize damage. There are, nevertheless, accidental spillages of oil at the terminal and refinery effluents contain oil though, because of the refining process used, less than at Fawley. All spilled oil is treated with chemical dispersants to reduce, if not eliminate, the amount coming ashore. Hydrocarbons have become incorporated in sediments (Fig. 3.13) but their distribution is related to the proportion of mud in the substratum rather than the sources of industry input. The sub-tidal benthic fauna remains rich throughout the area and is not obviously affected by this level of contamination.

IMPACT OF OFFSHORE OPERATIONS

An offshore platform may suffer a blow-out, resulting in an uncontrolled discharge of oil, but this is fortunately a rare event. Routine inputs of oil to the sea are from production water (see p. 31) and from small amounts of spilled oil washed off the platform by rain. Where oil-based drill muds (see p. 31) are used, bottom sediments receive a considerable amount of oil from contaminated cuttings.

No effect has been detected from oil discharged in production water or washed off platforms, and even the Ekofisk blow-out (see p. 32) had no detectable environmental impact.

Drill cuttings dumped on the seabed, however, have a profound effect on the benthic fauna. The drill cuttings blanket the seabed, creating anoxic conditions and the production of toxic sulphides in the bottom sediment with the almost total elimination of the benthic fauna. Surrounding this area, there is a recovery zone showing a succession of, first, opportunistic species such as *Capitella capitata*, then of species able to tolerate stressful conditions and to out-compete the less tolerant species, which gradually reappear further from the centre of disturbance (Fig. 3.14). This disruption of the biological system extends for about 500 m around the drilling platform in the North Sea oilfields (Fig. 3.15). In regions where bottom currents are stronger, it is likely that the discarded drill cuttings are more widely dispersed and there is less blanketing of the seabed.

It is difficult to separate the toxic effects of oil-based drill muds from those of sulphate-reducing bacteria which produce sulphides in areas where the seabed is blanketed with drill cuttings and organically enriched by the oil-based muds. Diesel degrades faster than the low-toxicity formulations, and platforms that have used diesel-based muds show greater enhancement of the opportunistic fauna than platforms that have used low-toxicity oil-based muds.

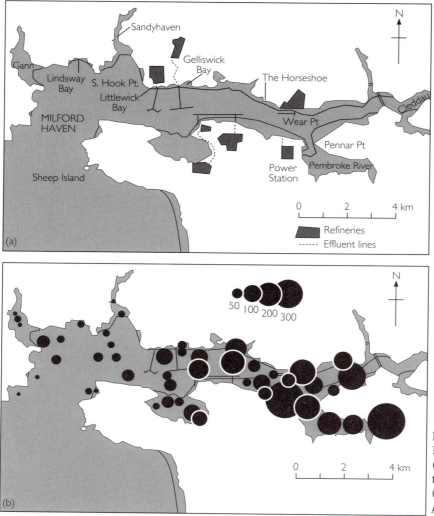

Fig. 3.13 Milford Haven, Pembrokeshire: (*a*) location of oil terminals and refineries, (*b*) hydrocarbons (in μg g^{-1}) in sediments.

PUBLIC HEALTH RISK FROM OIL POLLUTION

Some petroleum hydrocarbons are toxic to humans and there are a few cases on record of children being made seriously ill or even dying after inadvertently swallowing kerosene (paraffin). But humans have an extremely low taste threshold for petroleum hydrocarbons and the taste is particularly repulsive. There is therefore little risk of humans unknowingly receiving measurable doses of these toxins from contaminated food or drinking water.

Oil includes polycyclic aromatic hydro-carbons (PAH), some of which are known carcinogens. In the early 1970s there was a fear that PAH behaved in much the same way as chlorinated hydrocarbons like DDT—that they were resistant to bacterial attack and were excreted only slowly, if at all, by animals. As a result, it was concluded that these compounds might concentrate in the tissues of marine organisms with the concentrations increasing up the food chain to reach the highest levels in carnivorous fish. Human consumers of these fish might therefore be exposed to relatively large amounts of these carcinogens even in the absence of overt oil

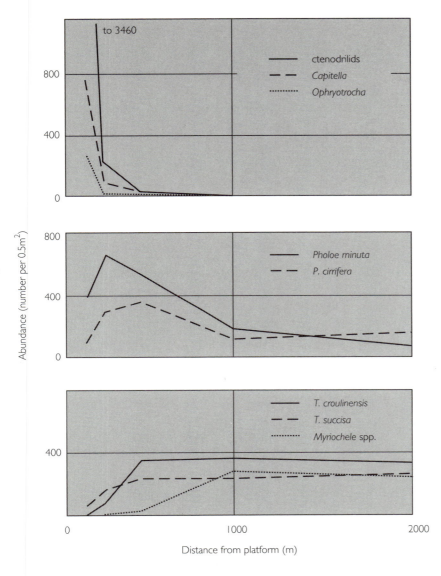

Fig. 3.14 Three types of abundance variation of benthic annelid worms influenced by disturbance at North Sea oil platforms.

pollution. Fortunately these fears have proved to be groundless. There is little evidence that petroleum hydrocrabons accumulate in marine organisms, and seafood contains low concentrations of PAH (as measured by benzo[*a*]pyrene, a potent carcinogen) (Table 3.4) compared with those occurring naturally in some other foodstuffs which are eaten in far greater quantity (Table 3.5). Consumption of seafood is unlikely to contribute more than 2–3 per cent of the normal dietary intake of PAH.

COMMERCIAL DAMAGE FROM OIL POLLUTION

Fisheries

Fixed installations where fish or shellfish are held in intensive mariculture are particularly vulnerable to damage from accidental oil pollution because the animals cannot escape. A slick of oil drifting through such an installation may inflict commercial damage quite incommensurate with the size of the spillage. Japan, where mariculture is more varied and

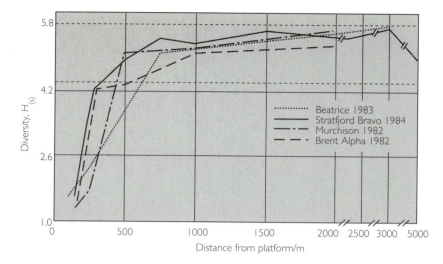

Fig. 3.15 Diversity of the benthic fauna (measured by the Shannon–Wiener index) in relation to distance from North Sea oil production platforms. Broken lines indicate the maximum and minimum values obtained in studies made before operations began.

Table 3.4 Concentration of benzo[*a*]pyrene (in μg kg^{-1}) in seafood from unpolluted and relatively polluted water

	Unpolluted	Polluted
Clam (*Mercenaria*)	0.38–1.1	8.2–16.0*
Shrimps	ND–0.5	ND–90.0
Crabs	ND–5.0	ND–30.0
Plaice	ND	0.05*
Herring	ND–0.1	0.4–13.0
Cod	ND–3.0	ND

ND = none detected.
*Evidence of tainting.

Table 3.5 Concentration of benzo[*a*]pyrene in (μg kg^{-1}) in various foodstuffs

	Benzo[*a*]pyrene content
Cooked meats and sausage	0.17–0.63
Cooked bacon	1.6–4.2
Smoked ham	0.02–14.6
Cooked fish	0.0
Smoked fish	0.3–60.0
Flour and bread	0.1–4.1
Vegetable oils and fats	0.4–36.0
Cabbage	12.8–24.5
Spinach	7.4

more advanced than in most other countries, has had many such unfortunate experiences.

In the open sea and, as a rule, in inshore waters, adult fish of commercial importance appear to be able to avoid areas affected by floating oil and are rarely killed. During the period when oil from a blow-out in the off-shore oilfield at Santa Barbara, California, contaminated coastal waters, spotter planes used by the local fishing fleets to detect fish shoals, found them in normal abundance in areas of clear water between the streamers of oil.

Since fish eggs and larvae are more sensitive than adults to toxins and are commonly in surface waters where they are likely to encounter high concentrations of petroleum hydrocarbons, they are expected to be particularly vulnerable to damage by oil pollution. In the laboratory, it has been found that Ekofisk crude oil reduced the hatching success of fertilized capelin eggs at concentrations of 10–25 nl l^{-1}. Concentrations of 250 μl l^{-1} cause developmental abnormalities in several species, which at best reduce their efficiency and in all probability result in their early death. Indeed, exposure of cod eggs to water-soluble extracts of Iranian crude oil for 100 hours caused immediate casualties, and these continued even after the larvae were transferred to uncontaminated water.

Many commercial species of fish and shell-fish produce enormous quantities of eggs, and even massive additional juvenile mortality may have no influence on the stock of adults which form the basis of the fishery.

Following the wreck of the *Torrey Canyon*, considerable areas of sea were sprayed with the very toxic dispersants then in use. A 90 per cent kill of pilchard (*Sarda pilchardus*) eggs was observed in the areas of spraying and a 50 per cent kill over a much wider area. Cornish pilchards form a small, isolated population but no shortfall could be detected two to three years later when the affected year-class were old enough to start appearing in the commercial catch.

One case is known in which an oil spill appears to have had a perceptible impact on fish stocks, and that was following the wreck of *Amoco Cadiz* on the Brittany coast in 1977 and the devastating pollution that resulted. The one-year-class of flatfish is, in this instance, thought to have been reduced.

On theoretical grounds, it has been calcu-lated that a major oil spill at the time and in the place of the main fish spawning in the North Sea would, at worst, cause a slight reduction in the overall catch for one year. In all probability, the effect would never be detected.

There have been reports of tumours and fin erosion in fish from areas chronically polluted by oil, and tumours and precancerous con-ditions in bivalves collected in such areas. Such pathological states occur naturally, but their incidence is increased in any polluted waters. They do not appear to be particularly related to petroleum hydrocarbons.

Most of the harmful effects of oil on fisheries refer to shellfisheries, either inter-tidally or in shallow water, and the damage may persist for years. A spillage of 700 t of diesel fuel (No. 2 fuel oil) in Buzzard's Bay, West Falmouth, Massachusetts in 1969, con-taminated shellfish beds, salt marshes, and beaches, and the oil became incorporated in the sub-littoral sediment. Some eight months after the accident, an area of 20 km^2 was polluted. Effects on crab populations were

still obvious seven years later. Also, following the grounding of the *Arrow* in Chedabucto Bay, Nova Scotia in 1970, 8000 t of Bunker C contaminated 240 km of shoreline. Popu-lations of the clam *Mya arenaria* were still adversely affected six years later.

Not all shellfish beds suffer in this way. A sudden decline in production of the Louisiana oyster fishery in 1932–3 coincided with the beginning of coastal oil extraction and gave rise to fears that the subsequent intensive development of inshore and offshore oilfields would be inimical to the oyster industry. In fact, this did not happen: the oyster fishery has had fluctuating success; occasional declines are not related to the activities of the oil industry but to other factors, such as incursions of freshwater into oyster lagoons. In fact, offshore oil platforms represent the only hard substrata in these waters and have attracted an exotic encrusting and sessile fauna; accompanying these are exotic fish species, and oil platforms have proved to be greatly favoured by sports fishermen in the area.

Tainting

As serious as losses resulting from deaths of fish and shellfish are, the most important commercial damage is from tainting. Light oils and the middle-boiling range of crude oil distillates are the most potent source of taint, but all crude oils, refined products, refinery effluents, wastes from petrochemical com-plexes, the exhaust from outboard motors burning oil–petrol mixture, and a host of other sources can impart an unpleasant flavour to fish and seafood which is detectable at extremely low levels of contamination. 'Oily' or 'petroleum' flavours are generally repulsive to humans and fish tainted in this way is unmarketable. Low levels of con-tamination may produce indefinite, but certainly detectable 'off-flavours' which are at least as damaging to the market.

The concentrations of oil in water neces-sary to cause tainting in finfish vary widely with the oil and the fish: fatty fish like salmon contract a taint much more readily than nonfatty species. In the natural environment, finfish rarely become tainted from floating oil because they avoid contaminated areas. The chief risk to commercial fishermen comes from fouling of fishing gear which may result in tainting of the whole catch. In chronically polluted waters, fish are either excluded or, generally, are not fished. Some chronically contaminated waters, such as Hong Kong harbour, are fished and the fish are tainted—the market in this case accepts tainted fish.

Shellfish may be affected in a variety of ways. Stranded oil contaminates the shells of molluscs without damaging them, but even a small quantity of oil on a few shells of mussels, cockles, or winkles can taint the whole catch if, as is commonly the case, they are boiled before eating. Since oil can be retained on shells for a considerable time, this source of tainting is very persistent.

A commoner form of shellfish tainting that affects all filter-feeders results from the uptake of fine droplets of emulsified oil in the course of normal feeding processes, and their subsequent incorporation into the tissues. Tainting in this way is common and bivalves are the principal subjects. The oily taste is lost within three to four weeks if the source of contamination is removed and the shellfish are transferred to uncontaminated waters.

The fishing industry is vulnerable to irrational moves in a way that few others are, and even the suspicion that fish may be tainted or contaminated is sufficient to depress the market. During the period when the wreck of the *Torrey Canyon* was in the news, fish sales on the Paris market fell by half even though no fish from any area affected by oil was ever on sale. There was a similar irrational response after the *Amoco Cadiz* wreck on the Brittany coast ten years later. Then, even cabbages grown in Brittany far from the coast were unmarketable for a period!

Tourism

Tourists prefer their bathing beaches free from oil and most coastal resorts put a good deal of effort into removing tar and oily residues from their amenity beaches. This oil pollution, mostly resulting from casual discharges, is a nuisance but not necessarily a heavy financial burden on the resort because the beaches are cleared of litter, plastics, and other debris regularly in the course of preserving the local amenities, and removing odd patches of tar adds little to the cost of beach cleaning.

Severe pollution resulting from a major accident in the vicinity of a resort is a different matter. It may well be beyond the means of the local community to deal with it and is generally treated as a national emergency calling for a national response. Tanker wrecks, like those of the *Torrey Canyon*, *Betelgeuse*, *Arrow*, and *Amoco Cadiz*, are in this category.

It is commonly said that oil pollution of beaches is harmful to the tourist industry, but although tourists may grumble there is little evidence that they actually stay away from beaches and resorts subject to minor oil pollution. Even large-scale pollution as from a tanker wreck does not apparently deter tourists who have been known to flock to view a disaster or potential disaster.

4

CONSERVATIVE POLLUTANTS

Organic wastes considered in previous chapters are subject to bacterial degradation, and environmental damage is caused by them when, by overloading the receiving waters, the intensity of bacterial action reduces the dissolved oxygen concentration, excluding fish and, at very low oxygen concentrations, other animals and plants as well. The seabed, lake bottom, or riverbed may become blanketed with fine material and undecayed organic matter leading to changes in, or even the elimination of, the benthic fauna. There may be a human health risk through the transmission of pathogens in untreated domestic sewage, but the chief hazard of organic wastes is the possibility of damage to natural resources.

Conservative pollutants present a different problem. They are not subject to bacterial attack or other breakdown, or if they are it is on such a long time-scale that for practical purposes they are permanent additions to the marine environment. Substances included in this category are:

(1) heavy metals such as mercury, cadmium, copper, zinc, and lead (considered in Chapter 5);

(2) halogenated hydrocarbons (considered in Chapter 6) which include such insecticides as DDT and dieldrin, and industrial chemicals in the group of polychlorinated biphenyls PCBs).

Plants and animals vary widely in their ability to regulate their metal content; most can do so only over a limited range. Both metals and halogenated hydrocarbons which cannot be excreted remain in the body in an unchanged state and are continually added to during the life of the organism. This is known as **bioaccumulation** and its dangers are illustrated by the following example. DDT has about the same toxicity to man as aspirin. A lethal dose of aspirin is about 100 tablets; the same quantity of DDT is also lethal. But it is possible to take 0.5–1.0 g of aspirin a day indefinitely without ill effect because it is excreted. DDT is not excreted, so lethal doses can be acquired after repeated exposure.

Animals feeding on bioaccumulators have a diet enriched in these conservative materials and if, as is commonly the case, they too are unable to excrete them or do so only slowly, they, in turn, acquire an even greater body-burden of the substance. This is **biomagnification**. Its chief significance is that top predators, which in the sea include man, may be exposed to very large concentrations of a conservative substance in their food. If the substances are toxic, they may cause damage at any level in the food chain, but the risks are obviously greatest for top predators. Conservative materials subject to bioaccumulation and biomagnification are therefore a potential human health risk as well as a threat to natural resources, and they have been responsible for human deaths. For this reason, conservative pollutants are regarded very seriously.

MEASURES OF CONTAMINATION

Concentrations of contaminants in organisms are expressed as:

$\mu g\, g^{-1}$ (parts per million, p.p.m.)
$\mu g\, kg^{-1}$ (parts per billion, p.p.b.)

(note that billion used in this context is 1000 million)

The concentrations may be calculated on the basis of the **wet weight** or **dry weight** of the tissue. The former is the weight of a sample of whole tissue removed from the body and often drained of free water; the latter is the weight after drying at 105°C to remove unbound water. If 80 per cent of the total weight of fish represents water, the dry weight concentration will appear as five times greater than the wet weight concentration. Tissues vary widely in their water content: feathers and hair contain practically none, lobster (*Homarus*) shell 25 per cent, but the muscle tissue of scallops (*Pecten*) is 75 per cent water; body fluids contain little solid matter and may be 90 per cent, or more, water. Dry weight concentrations provide the best basis for comparing the concentration of a substance in different tissues or different organisms, but many determinations are given only a wet weight basis. Unfortunately, it is not always made clear whether a concentration is based on wet or dry weight of the tissue and much confusion can result from this oversight.

For lipid-soluble substances (for example chlorinated hydrocarbons), it is often convenient and more realistic to quote the concentration as a proportion of the total lipid content of the body.

Concentrations of contaminants in sediments are always based on dry weight.

RELIABILITY OF ANALYSES

Chemical analysis has become very sophisticated and, with the increasing sensitivity of analytical techniques and equipment, it is now possible to measure substances at extremely low concentrations. Because of this, great precautions have to be taken to ensure that environmental samples are not contaminated in the course of collection or preparation for analysis. Many analyses, particularly earlier records, are suspect because of inadequate precaution against contamination from external sources. The analysis of lead, in particular, has suffered severe problems over a long period for this reason. Another problem is the loss of contaminants through adsorption on to containers or other materials employed in sample collection or pretreatment before analysis. Many analyses are also suspect because of the use of inappropriate analytical methods.

A laboratory may well produce consistent results when analysing replicate samples: the results show high **precision**. This is no guarantee of the **accuracy** of the analysis. In **intercalibration exercises** or **intercomparison trials** in which different laboratories independently analyse the same material, there is often a wide discrepancy between their results. For some compounds, notably polynuclear aromatic hydrocarbons, there is probably only a handful of laboratories around the world that can produce reliable results, although the same analytical equipment is universally available and in use.

The quality of analytical data can be controlled by the use of **reference materials**. These are samples of biological material or sediments available from a number of national and international agencies that contain a stated 'accepted value' of metals or other contaminants. If an appropriate reference material is analysed along with the environmental samples, it is possible to get a measure of the reliability of the results. A discrepancy between results of analysing the reference material and the environmental samples does not invalidate the latter; it merely gives guidance about the reliance that can be placed upon them.

TOXICITY

Toxicity is a measure of how poisonous a

substance is, or how large a dose is required to kill or damage an organism; the more toxic the substance, the smaller the lethal dose. Although the concept of toxicity appears straightforward, measuring it is subject to many complicating factors.

Measurement of toxicity

Organisms vary widely in the route by which toxins enter the body—via the mouth and digestive tract, at the gill surface, or across the integument—and toxins vary widely in the way they cause damage to the organism. Toxicity tests on aquatic animals generally avoid the problems of administering a known dose, or determining how much toxin an animal receives, by measuring the concentration of toxin dissolved in the water in which the organism is placed. In these circumstances, an organism is not given a single dose of toxin, but is exposed to it continuously, for the whole duration of the test.

In conducting toxicity tests, it is customary to expose a sample of test organisms to a particular concentration of toxin and measure how long it takes them to die. It may be difficult to detect with precision the moment of death and commonly some arbitrary criterion of lethal damage, such as immobility or loss of a defensive reflex, is used. Whatever criterion is used, because of variation in sensitivity to the toxin, the test organisms in the sample do not all die at the same time. Instead, mortality shows a sigmoid relationship to the period of exposure (Fig. 4.1). It is impractical and probably unrealistic to prolong an experiment until all the test organisms are dead, and the statistic used is the time for the death of 50 per cent of the test organisms, or the **median lethal time**, and is written as LT_{50} or LT_m.

The lethal time depends on the concentration of toxin to which an organism is exposed—the higher the concentration the shorter the time—and there may be a lower threshold concentration below which the material is not toxic. The LT_{50}, in itself, is therefore not a very useful statistic and it is usual to determine instead the concentration

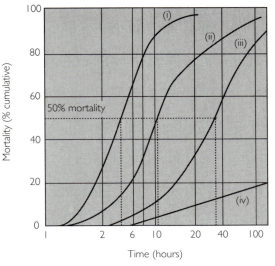

Fig. 4.1 Cumulative percentage mortality of mussels *Mytilus* in (i) 10^{-2} M, (ii) 10^{-3} M, (iii) 10^{-4} M, (iv) 10^{-5} M solution of zinc sulphate.

of toxin at which 50 per cent of the test organisms are killed within a specified time. This time is commonly 48 hours or 96 hours, though for tests on certain organisms, such as invertebrate larvae, it may be as short as 2 hours. The toxicity is then recorded as the **median lethal concentration**, written as **96 h LC_{50} or 96 h LC_m**, or with some other appropriate time indicated.

The median lethal concentration is measured by determining the median lethal time at several different concentrations, the latter usually being at logarithmic intervals, and reading the LC_{50} from a plot of the results (Fig. 4.2). More crudely, the mortality after 96 hours exposure to different concentrations may be recorded; the LC_{50} then lies between the concentrations which cause more and less than 50 per cent mortality.

When the test animals are subject to a single dose of the toxin, by injection or oral administration, the corresponding statistic is the size of dose that causes 50 per cent or median mortality: **LD_{50} or LD_m**. When a response other than death (or some arbitrary criterion of death) is used, EC, the effective concentra-

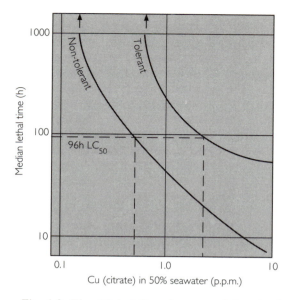

Fig. 4.2 The 96 h LC$_{50}$ of copper determined from the relationship between copper concentration (as citrate) in 50 per cent seawater and the median lethal time for copper-tolerant (from Restronguet Creek) and copper-intolerant (from the Avon estuary) *Nereis diversicolor*.

tion, or ED, effective dose, are used instead of LC or LD.

Some technical problems

The toxicity of metals depends on the form in which they are presented to the animal. Valency may change in seawater or complexes be formed between the metal and organic substances. Table 4.1 shows the form ('species') in which some heavy metals are believed to exist in the sea. Figure 4.3 shows the uptake of lead by the mussel *Mytilus edulis* when the metal is added as the nitrate, citrate, or complexed with organic molecules. Metal citrates are commonly used in toxicity tests because of the ready solubility of this salt, but lead citrate has three or four times the rate of uptake and final tissue concentration of the nitrate. Complexes with organic molecules double the rate of uptake. Organic complexes of heavy metals also tend to be more toxic than inorganic compounds: organomercurials and organotins are 10–100 times more toxic.

Organic arsenic, on the other hand, is less toxic than inorganic arsenical compounds.

Other toxins present different problems for designing realistic experiments. Oil, for example, is immiscible with water and toxicity is usually measured as that of the water-soluble components, although the concentrations and identity of these are not known in detail, and in any case, in the natural environment the organisms are likely to ingest emulsified droplets of whole oil. Furthermore, volatile components evaporate so that the nature of the material to which the organism is exposed changes throughout the course of an experiment.

As another example, chlorinated hydrocarbons have very low solubility in water and, in nature, animals are usually exposed to these compounds when they are adsorbed on to particles which the animal ingests. This is impossible to simulate accurately in the laboratory because of sedimentation of particles to the bottom of the container and adsorption of the chlorinated hydrocarbons on to the walls of the test chamber, which progressively reduces the availability of the material to the test organisms.

Organisms vary widely in their sensitivity to a toxin. Different species, of course, differ in their sensitivity, but even within a single species, sensitivity to a toxin depends on age, size, sex, reproductive condition, exposure to other stresses, nutritional state, and previous history, as well as the genetic constitution of the test organisms. The wide range of variation that may be due to these factors is illustrated in Table 4.2: young forms, for example, may be 100 times less sensitive or 1000 times more sensitive than adults of the same species.

Antagonism and synergy

In the natural environment, toxins are rarely present in isolation and they may interact with other substances. The combined effect of several toxins may be:

(1) the addition of one mortality to another;

(2) the containment of one within another;

Table 4.1 Concentration and speciation of trace metals in seawater

Metal		Area	Concentration ($\mu g\,l^{-1}$)	Main species in aerated water	
				35% salinity	10% salinity
Silver	Ag	N.E. Pacific	0.00004–0.0025	$AgCl_2^-$	
Aluminium	Al	N.E. Atlantic	0.162–0.864	$Al(OH)_4^-$, $Al(OH)_3$	
		N. Atlantic	0.218–0.674		
Arsenic	As	Atlantic	1.27–2.10	$HAsO_4^{2-}$	
Cadmium	Cd	N. Pacific	0.015–0.118	$CdCl_2$, $CdCl_3$, $CdCl^+$	Cd^{2+}, $CdCl^+$
		Sargasso Sea	0.0002–0.033		
		Arctic	0.015–0.025		
Cobalt	Co	N.E. Pacific	0.0014–0.007	$CoCO_3$, Co^{2+}	Co^{2+}, $CoCO_3$
Chromium	Cr	E. Pacific	0.057–0.234	CrO_4^{2-}, $NaCrO_4^-$	CrO_4^{2-}
Copper	Cu	Arctic	0.121–0.146	$CuCO_3$, Cu-organic	Cu-humic, $Cu(OH)_2$
		Sargasso Sea	0.076–0.108		
Iron	Fe	Arctic	0.067–0.553	$Fe(OH)_3$, $Fe(OH)_2^+$	
Mercury	Hg	N. Atlantic	0.001–0.004	$HgCl_4^{2-}$, $HgCl_3^-$	Hg-humic, $HgCl_2$
Manganese	Mn	Atlantic	0.027–0.165	Mn^{2+}, $MnCl^+$	Mn^{2+}
		Sargasso Sea	0.033–0.126		
Nickel	Ni	Arctic	0.205–0.241	$NiCO_3$, Ni^{2+}	Ni^{2+}, $NiCO_3$
		Sargasso Sea	0.135–0.334		
Lead	Pb	Central Pacific	0.001–0.014	$PbCO_3$, $PbOH^+$	
		Sargasso Sea	0.005–0.035		
Antimony	Sb	N. Pacific	0.092–0.141	$Sb(OH)_6^-$	$Sb(OH)_6^-$
Selenium	Se	Pacific and Indian	0.044–0.170	SeO_4^{2-}, SeO_3^{2-}	
Tin	Sn	N.E. Pacific	0.0003–0.0008	$SnO(OH)_3^-$	
Vanadium	V	N.E. Atlantic	0.83–1.57	HVO_4^{2-}, $H_2VO_4^-$	
Zinc	Zn	N. Pacific	0.007–0.64	Zn^{2+}, $ZnCl^+$	Zn^{2+}
		Sargasso Sea	0.004–0.098		
		Arctic	0.056–0.225		

(3) one may increase the mortality caused by the other (**synergy**);

(4) one may reduce the mortality caused by the other (**antagonism**).

Table 4.3 shows the result of an experiment in which the ciliate protozoan *Cristigera* was exposed to zinc, mercury, and lead salts in various concentrations and combinations of all three. The average percentage reduction in the growth of the culture was recorded in each case.

If the toxic effects are additive, the sum of the growth reduction for each of the metals acting alone should be the same as the reduction caused by all three acting together.

In this example, 0.005 p.p.m. mercuric chloride causes 12.1 per cent reduction, 0.3 p.p.m. lead nitrate reduces growth by 11.8 per cent, and 0.25 p.p.m. zinc sulphate reduces it by 14.2 per cent. The sum, $12.1 + 11.8 + 14.2 = 38.1$ per cent, should be the same as the effect of these concentrations of the salts in a mixture. In fact, as shown in the bottom right-hand figure in the table, it is 67.8 per cent.

In contrast to this synergistic effect, a mixture of 0.0025 p.p.m. mercuric chloride (causing 9.5 per cent reduction in growth if used alone) and 0.15 p.p.m. lead nitrate (causing 8.5 per cent reduction) produces only a 10.7 p.p.m. reduction in the growth of

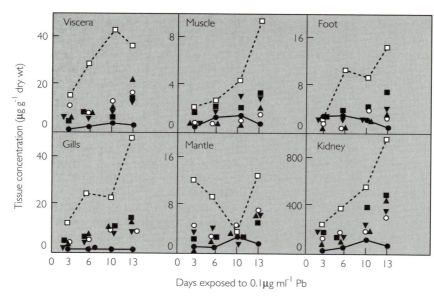

Fig. 4.3 Uptake of lead by various tissues of the mussel *Mytilus edulis*, with lead added as nitrate (○), citrate (□), humate (▲), alginate (▼), and pectinate (■), with seawater control (●).

Table 4.2 Sources of variation in bioassays

Source of variation	Ratio for comparing responses	Ratio range
Age of test organism	Old:young	0.01 to 1000
Sex of test organism	Male:female	0.5 to 5 or wider range
Genotypic difference	More resistant genotype:less resistant genotype	1 to 100 or wider range
Acclimation regime	Acclimated:non-acclimated	1 to 10 or wider range
Duration of exposure	Short bioassay test:long bioassay test	1 to 100

the population, instead of 9.5 + 8.5 = 18.0 per cent reduction expected if the effects were additive.

A variety of non-additive effects of multiple contaminants has come to light. Well-known examples are the antagonistic effects of selenium and mercury, and of zinc and cadmium, the former often reducing the effect of the latter in each case. Molybdenum in association with sulphate reduces the toxicity of copper to sheep; whether it confers the same benefit to marine animals is less certain.

Table 4.3 Percentage reduction of growth of *Cristigera* caused by different concentrations and combinations of metal salts

ZnSO₄ (p.p.m.)	0			0.125			0.25		
Pb(NO₃)₂ (p.p.m.)	0	0.15	0.3	0	0.15	0.3	0	0.15	0.3
HgCl₂ (p.p.m.)									
0	0	8.5	11.8	8.3	14.4	18.8	14.2	18.3	25.9
0.0025	9.5	10.7	14.5	13.9	16.2	22.6	18.8	29.0	51.3
0.005	12.1	18.7	21.8	18.9	21.3	23.2	35.5	36.5	67.8

Influence of other stresses

In the natural environment, organisms are exposed to a wide variety of stresses that either demand the expenditure of energy to counter them, or impair the organism's performance in some other way. Often, the responses to natural stresses may be regarded simply as part of the normal maintenance activities of the organism, as in osmotic or ionic regulation or any other physiological response to changes in the environment. Excretion of contaminants also places a demand on the energy resources of the organism, and its ability to make this defensive response to a toxin depends on what other calls there are on its energy resources. If demands are considerable, the organism will be the less able to survive exposure to the toxin.

Figure 4.4 shows how the resistance of the fiddler crab *Uca* to cadmium is affected by increased temperature and reduced salinity. The 240 h LC_{50} is about 48 p.p.m. when the crabs are under approximately normal conditions of temperature (10 °C) and salinity (30 per mille), but falls rapidly if the salinity is below 20 per mille. A combination of low salinity (10 per mille) and high temperature (30 °C) reduces the LC_{50} to only 2–3 p.p.m.

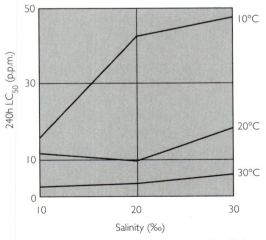

Fig. 4.4 Influence of temperature and salinity on the 240 h LC_{50} for cadmium to the fiddler crab *Uca pugilator*.

Delayed impact of toxins

Standardized toxicity tests measure only the mortality that takes place during a test. It is recognized, of course, that the longer the test goes on, the greater the number of test organisms that will die, but the test takes no account of mortality that may take place after organisms that have been exposed to a toxin are removed to uncontaminated water. A toxin that has a delayed effect of this kind may well have a deceptively low toxicity rating in short-term tests.

Figure 4.5 shows the survival of mussels, *Mytilus edulis*, living in uncontaminated seawater for a period of three weeks following 24 hours exposure to copper sulphate in various concentrations. Concentrations up to 0.3 p.p.m. have little impact. Higher concentrations cause deaths, but these do not appear until after a time-lag of five or six days.

Sub-lethal effects

It is often possible to detect damage to organisms by toxins at far lower concentrations that those that kill them. These sub-lethal responses range from little more than physiological adjustment to changed circumstances, to major physiological stress or developmental abnormalities which, in the natural environment, would be likely to result in early death.

Erosion of fins and the development of pre-cancerous growths (papillomas) on the ventral surface are commonly observed in a proportion of flatfish living on very contaminated sediments. Since the same effect has been observed whether the contaminant is sewage sludge, oil, or titanium dioxide waste, the response is not specific to a particular pollutant.

The ingestion of crude oil by herring gulls (*Larus argentatus*) and some other sea birds causes damage to the intestine and liver, and impairs the functioning of the nasal salt glands.

Skeletal deformities are relatively common in fish, particularly from polluted waters: as much as 8.4 per cent of herring caught in the

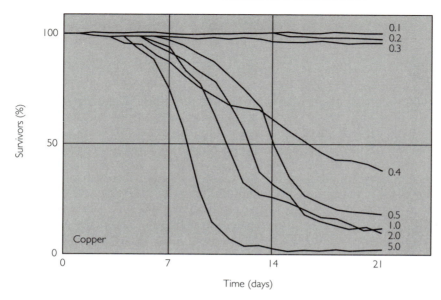

Fig. 4.5 Survival of mussels, *Mytilus edulis,* during three weeks following exposure to various concentrations (in p.p.m.) of copper sulphate.

southern North Sea in a single year are afflicted in this way, and 6 per cent of coregonid white fish caught in the German Bodensee have spinal deformities. To some extent this condition is due to hereditary factors, parasitic infection, developmental errors, or other natural causes, but a variety of toxins, including chlorinated hydrocarbons and organophosphorus pesticides, dispersed oil, and heavy metals, can also cause it.

Sub-lethal concentrations of 0.05–0.10 p.p.m. copper sulphate and zinc sulphate cause the production of abnormal larvae in the polychaete *Capitella capitata*. These larvae are bifurcated (Fig. 4.6) and fail to survive beyond the eight-segment stage. A similar abnormality occurs in 10–16 per cent of larvae of this species in the second generation reared in water contaminated with a detergent.

Physiological stress is not necessarily harmful. Commercial fish hatcheries often subject fry to as many as possible of the stresses they are likely to encounter in nature in order to increase their survival rate after release. In theory, such adaptive physiological responses are distinguished from harmful ones if they contribute to the survival, growth,

and reproduction of the species, but the distinction is not always easy to make.

Interpretation of toxicity data

Toxicity, as measured in the laboratory, has limited relevance to events in a natural situation (though it is very significant within those limits), and the results of toxicity tests must be interpreted with caution. A variety of factors relating to the toxin, the test organism, and the design of the toxicity test may all influence the results. Even when these factors can be adequately controlled, a different range of factors relating to the population dynamics and ecological role of the species have to be taken into account before the results of laboratory tests can be translated into terms applicable to the natural environment.

Table 4.4 summarizes the factors influencing the toxicity of heavy metals in solution. It may be possible to standardize the organisms used in tests and, by suitable control of experimental conditions, to control the form in which a metal is presented to the organisms. The use of continuous-flow rather than static systems avoids depletion of oxygen and accumulation of excretory products, or loss of the toxin in the course of a prolonged toxicity

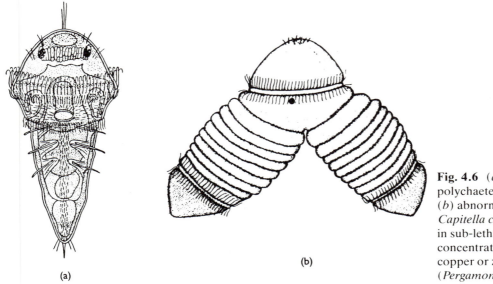

Fig. 4.6 (*a*) Normal polychaete larva; (*b*) abnormal larva of *Capitella capitata* living in sub-lethal concentrations of copper or zinc sulphate. (*Pergamon Press.*)

(a)

(b)

Table 4.4 Factors influencing the toxicity of heavy metals in solution

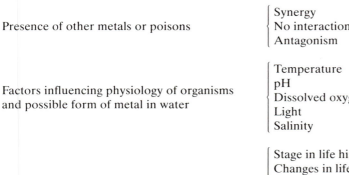

Form of metal in water	Inorganic Organic	Soluble	Ion Complex ion Chelate ion Molecule
		Particulate	Colloidal Precipitate Adsorbed
Presence of other metals or poisons	Synergy No interaction Antagonism	More-than-additive Additive Less-than-additive	
Factors influencing physiology of organisms and possible form of metal in water	Temperature pH Dissolved oxygen Light Salinity		
Condition of organism	Stage in life history (egg, larva, etc.) Changes in life cycle (e.g. moulting, reproduction) Age and size Sex Starvation Activity Additional protection (e.g. shell) Adaptation to metals		
Behavioural response	Altered behaviour		

test. Even with such refinements, however, the test situation is far removed from that in the natural environment where the behaviour of the toxin may be modified by the physico-chemical environment or because it interacts with other substances that are also present.

Further to these complications, the ecological consequences of mortality caused by a toxin, which are a principal reason for measuring toxicity, introduce many more questions. Whether or not a local mortality affects the local population depends upon the population dynamics of the species (the survivors may flourish through reduced competition), the mobility of the species (adjacent populations may fill gaps caused by the mortality), and the persistence of the toxin. Each species is a member of a community and interacts with other members, all of which are exposed to the toxin. Population changes among them may secondarily affect the first species. These factors are considered in Chapter 10.

Toxicity tests have been conducted on few species and are focused particularly on those that are hardy, readily available, and lend themselves to laboratory investigation. These tests are at least useful in ranking substances in order of toxicity to a spectrum of test organisms. There will undoubtedly prove to be organisms unusually sensitive or unexpectedly more resistant to the toxin than the standard tests reveal, but they provide a method of screening new products and giving early warning of more obvious hazards. With suitably refined techniques, it may be possible to identify which of several contaminants is primarily responsible for known damage or to identify which contaminants in combination are causing ecosystem damage. In general, however, these tests are not useful in predicting the ecological impact of a toxin.

ACUTE AND CHRONIC EXPOSURE

It is important to distinguish between two ways in which an organism may be exposed to a poisonous substance. **Acute** exposure is a single event, **chronic** exposure continues over a long time. The wreck of an oil tanker on the coast is an acute incident: a single, damaging event from which, once the oil is removed, recovery can begin. A refinery effluent containing petroleum hydrocarbons is chronic: the initial damage may be much less than from an acute oil spill, but since the discharge continues, recovery cannot begin and, indeed, the damage is progressive.

In the same way, the **acute toxicity** of a substance is the lethal amount when it is administered in a single dose. The **chronic toxicity** is the lethal amount when the substance is administered continuously over a long period. It generally follows that the LD_m for acute poisoning is much higher than that for chronic poisoning.

DETOXICATION

Extraordinarily high concentrations of heavy metals have been found in some animals in the sea: 2000 p.p.m. (dry weight) of cadmium in the digestive gland of some scallops and 57 000 p.p.m. (dry weight) of zinc in some oysters. If such concentrations were circulating freely in the animal they would be lethal, but in at least some animals methods have been developed for storing metals in the body in an innocuous form. Two methods of such **detoxication** are known.

One method involves the formation of **metallothioneins**. Exposure of the animal to a low concentration of metal induces the synthesis of low-molecular-weight proteins which form a complex with the metal. The metal is then unable to become involved in chemical processes, interference with which is the source of its toxicity. This phenomenon was first discovered in mammals but very similar metal-binding proteins have been found in a variety of other animals and they are probably widespread. Cadmium-binding proteins have been identified in seals, rockfish (*Sebastodes*), and molluscs. Both copper and cadmium are bound by a thionein in crabs, where the primary storage site of the metallothionein is the hepatopancreas. Metallothioneins incorporating zinc, copper, and cadmium have been found in the sea urchin *Strongylocentrotus purpureus*. Metallothioneins binding gold, silver, and mercury have been found in the rat liver and it is possible that these and perhaps others will prove to occur in a variety of marine animals.

A second method of detoxicating metals has now been detected in nearly all animal groups. The metal is stored in granular form; the granules are bounded by a membrane and are therefore isolated from chemical activity in the cell. The membrane may be derived from endoplasmic reticulum, the Golgi apparatus, or the cell membrane of the storage cell. So far, three types of granule have been found:

(1) copper-containing granules which generally also contain sulphur and perhaps traces of calcium;

(2) calcium-containing granules either of high purity, or in which, in addition to calcium, there is also manganese, magnesium, and phosphorus, and sometimes other metals such as zinc, cadmium, lead, and iron;

(3) iron-containing granules.

The calcium-containing granules of high purity are commonly involved in some essential activity in the animal and are not concerned with heavy-metal detoxication. Examples of such granules occur in the calciferous glands of earthworms, and they are stored and used for calcification of the shell in bivalves and barnacles.

Impure calcium-containing granules appear to be a system for incorporating and detoxifying metals, though the mechanisms of this are not yet fully understood.

Iron appears to be accumulated through life by the heart urchin *Bryssopsis*, but in many organisms the metal granules can be excreted, though this may involve a special process which is not yet understood. Reduction of the amount of copper and zinc in the commercial shrimp *Penaeus semiculcatus* during pre- and post-moult periods suggests that the granules are eliminated from the hepatopancreas at these times.

Some of these detoxicating mechanisms have only a limited capacity, though little study has been made of this feature of them. When the maximum capacity of the detoxicating mechanism is reached, the animal is then exposed to the toxic effects of the heavy metal.

5

METALS

INPUT ROUTES

Metals are natural constituents of seawater (see Table 4.1) and assessing the effect of inputs resulting from human activities is complicated by the very large natural inputs from the erosion of ore-bearing rocks, wind-blown dust, volcanic activity, forest fires, and vegetation.

Most rivers make a major contribution of metals to the sea, the nature of the input depending on the occurrence of metal and ore-bearing deposits in the drainage area. Where the river passes through urban or industrialized centres, the metal burden is augmented by human wastes and discharges.

The intense sedimentation in estuaries (see p. 11) traps a large quantity of metals which become adsorbed on to sediment particles and carried to the bottom. Sediments in industrialized estuaries with major ports contain the legacy of a century or more of waste discharges. Regular dredging of shipping channels in such areas produces large quantities of dredging spoil, heavily contaminated with metals, which is usually dumped at sea.

A third major input of metals to the sea, particularly near the coasts of industrialized countries, is from the atmosphere (see below). Atmospheric processes are still poorly understood and estimates of inputs of metals to the sea by this route vary quite widely, but all estimates suggest that this input is considerable.

Much smaller quantities of metals are added to the sea by direct discharges of industrial and other wastes by pipeline, and in sewage sludge and industrial wastes that are dumped at sea. Ocean incineration (see p. 86) makes a very small contribution. Although these inputs are relatively small, they may be locally significant if they are added to areas of sea where there is limited water circulation.

Table 5.1 shows the estimated inputs (in 1987) of four metals to the North Sea by various routes. The atmosphere is a more significant route for lead than it is for other metals but, apart from this, the relative importance of the various routes is much the same for all metals. The North Sea is a relatively enclosed body of water, surrounded by industrialized countries, with the rivers entering it draining much of the most heavily industrialized parts of central and western Europe. The figures in the table therefore reflect an extreme situation.

ATMOSPHERIC INPUTS

An important route by which some metals enter the sea is via the atmosphere. There are large natural inputs of some metals, such as aluminium in wind-blown dust derived from rocks and shales, and mercury from volcanic activity and degassing of the earth's crust, but for some metals, inputs to the atmosphere as a result of human activities are greater, sometimes much greater, than natural inputs (Table 5.2).

These atmospheric contaminants may exist as gases (mercury, selenium, and boron) or aerosols (most other metals). They are

Table 5.1 Input of four metals to the North Sea (t yr^{-1})

Source	Copper	Mercury	Lead	Zinc
Rivers	1290–1330	20–21	920–980	7360–7370
Dredgings	1000	17	170	8000
Atmosphere	400–1600	10–30	2600–7400	4900–11 000
Direct discharge	315	5	170	1170
Industrial dumping	160	0.2	200	450
Sewage sludge	100	0.6	100	220
Incineration	3	—	2	12

Table 5.2 World-wide emissions of trace metals to the atmosphere (in thousand t yr^{-1})

Metal	Natural sources	Anthropogenic sources
Arsenic	7.8	24
Cadmium	0.96	7.3
Copper	19	56
Nickel	26	47
Lead	19	449
Selenium	0.4	1.1
Zinc	4	314

deposited by **gas exchange** at the sea surface or the fallout of particles (**dry deposition**) or scavenged from much of the air column by precipitation within clouds (**wet deposition**). Sometimes the particles provide the nuclei about which water drops form. The length of time that a contaminant remains in the atmosphere, and so the distance it travels in an air mass, depends on its reactivity if it is a gas or its density if a particle. Lead has a residence time of about 5 days; gaseous organic compounds, tens or hundreds of days.

It is difficult to estimate atmospheric inputs to the sea; rates of precipitation over the sea, crucial for estimating wet deposition for example, are very poorly known. Table 5.3 shows the 1986 estimates of metal inputs from the atmosphere to various sea areas, but these figures may need revision as more information becomes available.

Air–sea interactions are not a one-way process. Bubbles bursting at the sea surface release sea salt particles to the atmosphere and there is evidence that these particles become enriched with other contaminants during their formation. The sea is therefore a **source** of contaminants to the atmosphere as well as a **sink** for atmospheric contaminants. Too little is understood about these processes to make very firm estimates of the relative importance of inputs to and outputs from the

Table 5.3 Transfer of metals from the atmosphere to the sea surface (in ng cm^{-2} yr^{-1})

Element	North Sea	Western Mediterranean	South Atlantic Bight	Tropical North Atlantic	Tropical North Pacific
Aluminium	30 000	5000	2900	5000	1200
Manganese	920	—	60	70	9
Iron	25 500	5100	5900	3200	560
Nickel	260	—	390	20	—
Copper	1300	96	220	25	8.9
Zinc	8950	1080	750	130	67
Arsenic	280	54	45	—	—
Cadmium	43	13	9	5	0.35
Mercury	—	5	24	2.1	—
Lead	2650	1050	660	310	7.0

sea surface. Some idea of the possible scale of natural contributions from the sea to the atmosphere can be gained from the discovery that dimethyl sulphide synthesized by phytoplankton blooms of coccolithophores in the North Sea and released to the atmosphere, may be responsible for half the acid rain falling on coastal areas in summer.

UPTAKE OF METALS

Many metals are essential for living organisms:

(1) the respiratory pigment haemoglobin found in vertebrates and many invertebrates contains iron;

(2) the respiratory pigment of many molluscs and higher crustaceans, haemocyanin, contains copper;

(3) the respiratory pigment of tunicates contains vanadium;

(4) many enzymes contain zinc;

(5) vitamin B_{12} enzymes contain cobalt.

Metals of biological concern may be divided into three groups:

(1) light metals (sodium, potassium, calcium, and so on) normally transported as mobile cations in aqueous solutions;

(2) transitional metals (for example iron, copper, cobalt, and manganese) which are essential in low concentrations but may be toxic in high concentrations;

(3) metalloids (such as mercury, lead, tin, selenium, and arsenic) which are generally not required for metabolic activity and are toxic to the cell at quite low concentrations.

Transitional metals and metalloids are usually collectively termed heavy metals.

Particularly among polychaete worms, metals are accumulated for unexpected reasons. High concentrations of zinc in the jaws of nereids and copper in the jaws of *Glycera* have a structural role, strengthening the tips of the jaws. *Nereis* also accumulates

manganese in its jaws. Copper accumulated in the gills of the ampharetid *Melinna palmata* has the defensive function of reducing the palatability of the worm to predators. The cirratulid *Tharyx marioni* has a high and constant level of arsenic in its feeding palps; its function is unknown.

Absorption of heavy metals from solutions is dependent on active transport systems in some microorganisms and sea urchin larvae, but generally in plants and animals it is by passive diffusion across gradients created by adsorption at the surface, and by binding by constituents of the surface cells, body fluids, and so on. An alternative and important pathway for animals is when metals are adsorbed on to or are present in food, and by the collection of particulate or colloidal metal by a food-collecting mechanism such as the bivalve gill. There is considerable variation in the extent to which plants and animals can regulate the concentrations of metals in the body: plants and bivalve molluscs are poor regulators of heavy metals; decapod crustaceans and fish are generally able to regulate essential metals such as zinc and copper, but non-essentials such as mercury and cadmium are less well regulated.

MERCURY

Sources and inputs to the marine environment

The annual world production of mercury reached a peak of 10 600 t in 1971, falling to a little over 6000 t in 1987. A typical pattern of the use of mercury in an industrialized country is shown in Table 5.4.

A substantial proportion of this annual production eventually reaches the natural environment. Table 5.5 shows the estimated losses of mercury in Finland and Sweden during the 1950s and 1960s. About half the mercury lost was from the chlor-alkali industry. Chlorine and caustic soda are manufactured electrolytically using mercury electrodes and it was common in this process for 150–200 g of mercury per tonne of

Table 5.4 Consumption (t) of mercury in the United States

	1968	1974–5	1984
Electrical apparatus	667.0	783.4	1170.0
Chlor-alkali industry	602.0	788.8	253.0
Paints (including anti-fouling)	369.0	370.0	160.0
Industrial and control instruments	275.0	319.9	98.0
Dental	106.0	131.1	49.0
Agriculture	118.3	91.4	—
Catalysts	66.0	81.8	112.0
Laboratory use	69.0	71.6	7.5
Pharmaceuticals	15.0	22.4	—
Paper and pulp	14.0	8.6	—
Amalgams	9.0	8.6	—
Other	298.0	205.6	48.0
Total	2628.0	2882.2	1797.5

product to be lost to the atmosphere or in waste water. Most of the other half of the 'lost' mercury came from uses which inevitably disperse mercurial compounds into the environment: 'slimicides' for use in the lumber and paper pulp industries to prevent fungal growth, antifouling paint for ships' hulls, pesticides and seed dressings in agriculture, and in pharmaceuticals.

These inputs were estimated to amount to about 5000 t yr^{-1} world wide, but a further 3000 t yr^{-1} is derived from burning fossil fuels. Coal and oil contain only traces of mercury, varying widely but perhaps averaging 1 μg g^{-1} (1 p.p.m.), but they are burned in such large quantities that their input to the atmosphere is considerable.

Following the discovery in the early 1960s of the dangers to human health of mercury in

Table 5.5 Consumption of mercury (t yr^{-1}) in Finland and Sweden, and estimated losses to rivers, lakes, and the sea

	Consumption (t yr^{-1})	
	Finland 1954–66	Sweden 1967
Chlorine industry, increase of capacity	22	?
Chlorine production, replacement for losses	10	25–38
Paper and pulp industry, as slimicides and fungicides	10	14
Agriculture, as pesticides	4	5
Other uses	4	~ 30
Total	50	74–87
	Losses into waters (t yr^{-1})	
	Finland 1954–66	Sweden 1967
Chlorine industry	4	8
Paper and pulp industry		
As slimicides	3.6	3.2*
As caustic soda	0.5	1.0
Agriculture, as pesticides	0.3	0.3
Other industries	0.2	1.5
Total	8.6	14.0

*1941–68.

the sea, there has been a steady reduction in inputs from major sources. Current EC regulations limit the discharge of mercury in waste water from the chlor-alkali industry to 5 g per tonne of product, and similar standards have been adopted in a number of other parts of the world. Some former uses of mercurial compounds have been phased out altogether. From Table 5.4 it can be seen that mercury is no longer used in agricultural pesticides, pharmaceuticals, or by the timber and paper industry in the United States. Mercury is prohibited in antifouling paints in many countries, including the EC.

These changes to practice have not been trouble-free. Sweden banned the use of mercurial slimicides in the timber and pulp industry in 1967 and other countries followed suit. However, pentachlorophenol, which replaced the mercurial slimicides, proves to be damaging to marine macrofauna at concentrations of 75 μg l^{-1} and accumulates in at least one marine organism, the polychaete *Lanice*. The replacement of mercurial by tin compounds in antifouling paints also proved damaging (see p. 80) and now the use of tin-based antifouling paints has had to be restricted.

Mercury releases do not originate only from industrialized countries in the northern hemisphere. In the Amazon area, mercury is used to recover gold and silver and results in 70 t yr^{-1} of mercury emissions to the atmosphere as well as serious contamination of river fish downstream of the operations.

Natural inputs of mercury are from two sources. About 3500 t yr^{-1} is derived from the weathering of mercury-bearing rocks, but anything from 25 000 to 150 000 t yr^{-1} is released to the atmosphere as gases from volcanic areas, geothermal vents, and other sites of degassing of the Earth's crust.

Although it cannot be estimated reliably, atmospheric input is important in the open oceans where mercury concentrations vary between 0.001 and 0.005 μg l^{-1} (p.p.b.). Major sources of mercury in coastal waters are rivers, marine outfalls, and wastes dumped directly into the sea. The existence of such localized inputs is reflected in the raised mercury content of local plankton and mussels.

Dissolved mercury in the sea is in the form of $HgCl_4^{2-}$ or $HgCl_3^-$ but, to a considerable extent, mercury is adsorbed on to particulate matter and is not in solution. It also forms stable complexes with organic compounds occurring in the sea, especially sulphur-containing proteins or, in waters of reduced salinity, humic substances (Table 4.1, p. 57). In anoxic substrata, mercury may be present as Hg, HgS, and HgS_2. Microbial systems in the sea can convert all these inorganic forms of mercury into methyl mercury, which is readily released from sedimentary particles into the water and may then be accumulated by living organisms.

Mercury in algae and invertebrates

Organic forms of mercury are more toxic than inorganic salts. Table 5.6 shows the 18 h LC$_{50}$ of a number of mercury compounds to the red alga *Plumaria elegans*, and there is a similar pattern of toxicity of these compounds to larvae of the barnacle *Elminius* and brine shrimp *Artemia*. As with other metals, bivalve molluscs take up mercury from the surrounding water very quickly. Figure 5.1 shows the uptake of mercury by the commercial species *Crassostrea virginica* exposed to mercuric acetate. The distribution of

Table 5.6 Toxicity of various mercury compounds to the red alga, *Plumaria elegans*

Compounds	18 h LC$_{50}$ (p.p.b. Hg)
Methyl mercuric chloride	44
Ethyl mercuric chloride	26
n-Propyl mercuric chloride	13
n-Butyl mercuric chloride	13
n-Amyl mercuric chloride	13
Isopropyl mercuric chloride	28
Isoamyl mercuric chloride	19
Phenyl mercuric chloride	54
Phenyl mercuric iodide	104
Mercuric iodide	156
Mercuric chloride	3120

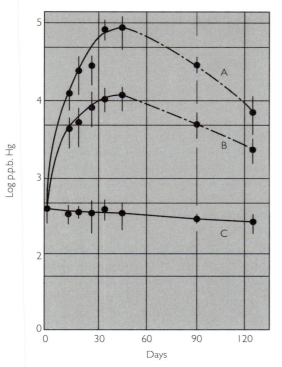

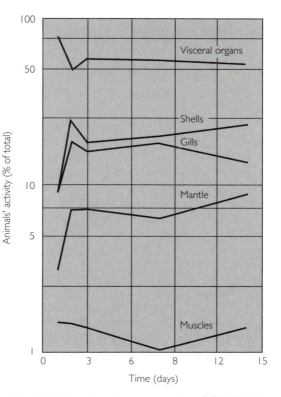

Fig. 5.1 Uptake of mercury by the bivalve *Crassostrea virginica* exposed to (A) 100 p.p.b. and (B) 10 p.p.b. mercuric acetate for 45 days (————), and subsequent loss of mercury after transfer to uncontaminated seawater (– · – · – · –). (C) Control.

Fig. 5.2 Uptake of mercury (as $^{203}HgCl_2$) in different tissues of the bivalve *Venerupis decussata*. (© *FAO*.)

mercury in the tissues varies (Fig. 5.2) and in bivalves the highest concentration is in the viscera, the lowest in the muscles. Once the molluscs are returned to uncontaminated water, the mercury is lost from the body.

Mercury in fish

Most species of fish in oceanic waters contain 150 μg kg^{-1} (0.15 p.p.m.) mercury in muscles, though much higher values are found in fish from contaminated waters. Cod taken from the Sound between Denmark and Sweden, which is heavily contaminated with mercury, contain up to 1.29 p.p.m.; those from the North Sea 0.15–0.20 p.p.m.; and those from Greenland only 0.01–0.04 p.p.m. Some species, notably tunny (*Thunnus* spp.), swordfish (*Xiphias gladius*), marlin (*Makaira indica*), and some other large, oceanic pelagic

fish, naturally contain high concentrations of mercury. Concentrations of 1 p.p.m. in the muscle are common and may be as high as 4.9 p.p.m. There are several reasons for this. These fish are large carnivores at the end of food chains and their diets therefore contain high levels of mercury resulting from bio-accumulation and biomagnification. Food is not their only source of mercury. They are very active fish with a high metabolic rate. They swim continuously with their mouths open, so producing a forced flow of water across the gills. This results in a large uptake of oxygen, but also of metals (including mercury) dissolved in the water. Much of this mercury is in the form of methyl mercury and, because the fish cannot excrete it, its concentration in the tissues increases with the age of the fish. It is the large, old specimens that have high body burdens of mercury, and warnings are given about the danger of eating the flesh of very large fish of these species.

Halibut (*Hippoglossus hippoglossus*) is another long-lived species that accumulates high concentrations of mercury. A big halibut of 300 kg may be 50 years old. All specimens over 115 kg and half those weighing more than 60 kg may be expected to contain over 1 p.p.m. muscle mercury. As with most fish, more than 90 per cent of this is in the form of methyl mercury and such fish are unsuitable for human consumption. They may be used for fishmeal, where their high concentration of mercury is diluted by the large mass of small fish containing little mercury that make up the bulk of the fishmeal used for animal feed.

Mercury in birds

Birds tend to accumulate mercury in their liver and the feathers. Most published records refer to liver or feather mercury and measurements are given on the basis of dry weight (see p. 54). Levels of contamination therefore appear far higher than those in fish for which measurements generally refer to edible parts (muscle mercury) and are given on a wet weight basis. The fact that mercury accumulates in feathers is interesting because it allows a study to be made of museum specimens, revealing levels of mercury contamination at various places and times in the past when the specimens were collected.

The larger sea birds may live for 20 years or more and those feeding at high trophic levels are likely to accumulate mercury. Species that feed on or close to shore are most at risk: black guillemot (*Cepphus grylle*) are inshore feeders and contain more mercury than common guillemot (*Uria aalge*) and Brünnich's guillemot (*U. lomvia*), which feed out at sea.

The contamination of the Baltic is reflected in the mercury levels in fish-eating birds from Scandinavia. Wing feathers of ospreys (*Pandion haliaetus*) replaced there contain more mercury than those replaced in their winter quarters in Africa. Feathers of Baltic guillemots contain three to five times as much

mercury as those from Faeroese or Greenland birds. Measurements of the mercury in the feathers of museum specimens of Baltic Guillemots has shown that the level of contamination rose by two to five times in the first half of this century, though after the late 1960s it started to decline again.

The transfer of mercury to feathers which are periodically moulted may be an important safety valve for birds exposed to heavy-metal contamination. In Sweden and Finland, the population of both ospreys and white-tailed eagles (*Haliaetus albicilla*) has declined during this century and mercury has been thought to be a factor. These are fish-eating birds and the mercury concentration in feathers of the white-tailed eagle increased from an average of 6.6 p.p.m. in Swedish museum specimens collected between 1880 and 1940, to about 50 p.p.m. in 1964–5. In the mid-1960s, white-tailed eagles are known to have died from mercury poisoning in southwest Finland, but ospreys appear not to have suffered in this way.

A study of osprey nestlings in Finland showed that there is certainly a relationship between the concentration of mercury in their diet and that in their plumage, although there is a limit to the concentration of mercury that can be acquired by feathers (Fig. 5.3). Even so, in 2.5 per cent of the nestlings examined, the mercury content of the feathers was over 30 p.p.m. Comparisons of nestlings from Lapland (uncontaminated) and from Hämeen-kyrö in southwest Finland (relatively contaminated) showed that in the former the concentration of mercury in the liver (dry weight) was twice that in the feathers; in the contaminated area, the feather concentration was actually greater than that in the liver (Table 5.7). More striking, in terms of the total body burden of mercury, nestlings containing 60 μg of mercury have one-third of it in the plumage, but nestlings containing 770 μg stored two-thirds of it in the plumage. It is not known if white-tailed eagles have less ability to transfer mercury to the plumage, but in ospreys this appears to be a method of removing mercury from tissues where it can

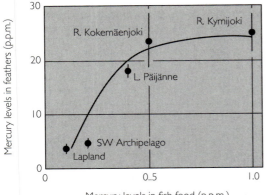

Fig. 5.3 Mercury concentration in feathers of nestling ospreys in five areas of Finland in relation to the mercury concentration in local fish.

Table 5.7 Mercury in osprey nestlings from Lapland (uncontaminated) and southwest Finland (relatively contaminated)

	Inari (Lapland)	Hämeenkyrö (SW Finland)
Liver (wet wt) (p.p.m.)	0.2	1.8
(dry wt) (p.p.m.)	0.8	7.1
Feathers (p.p.m.)	0.4	8.3
Total body burden		
(μg)	60.5	772.8
% in plumage	32.6	65.5

interfere with metabolic processes. This is, in fact, a method of detoxication and excretion.

Mercury in sea mammals

Marine mammals accumulate large quantities of mercury from their food apparently without coming to harm. Selenium antagonizes the toxic effects of mercury, and in seals, sea lions, and dolphins, the selenium concentration in the tissues keeps pace with the amount of mercury. Mercury selenide found in the connective tissue of the liver of dolphins is apparently a product of a detoxifying mechanism for the methyl mercury acquired from their food.

Minamata disease

That mercury is toxic to humans has been known for centuries. Mad hatters in Victorian England got their name from convulsions and loss of neuromuscular coordination, symptomatic of chronic poisoning by the mercury used in the treatment of felt in hat manufacture. In this century, mercury has been the only contaminant introduced by man into the sea, apart from pathogens in sewage, that has certainly been responsible for human deaths.

Mercury poisoning resulting from pollution of the sea occurred at the small Japanese coastal town of Minamata where part of the population was dependent on fishing for a livelihood and the only industry was a factory which began producing vinyl chloride and acetaldehyde in 1952. This illness affected only fishermen and their families and first appeared in 1953, but was not diagnosed as metal poisoning derived from fish and seafood taken from part of Minamata Bay until 1956. In all, 2000 cases were recognized; of these, 43 died during the epidemic and over 700 of the survivors were left with severe permanent disabilities. Fishing in part of the bay was banned from the start of 1957 and the epidemic halted. It was not until 1959 that it was shown that mercury was the toxic element involved, and 1960 that the source was the effluent from the vinyl chloride factory discharged into the bay near the affected community.

A second outbreak of mercury poisoning in Japan occurred among fishermen living near the mouth of the Agamo river in 1965. Like the Minamata tragedy, this was caused by the contamination of fish by mercury from an industrial effluent, in this case, from a factory 60 km upstream.

The Minamata disaster was investigated in detail. It proved to be a classical example of methyl mercury poisoning. Methyl mercury is more dangerous to humans than inorganic mercury because it cannot be excreted. It therefore acts as a cumulative poison and, since it can cross the barrier from blood vessels in the brain into the nervous tissue (the blood–brain barrier), it causes progressive and irreversible brain damage. The chief danger from inorganic mercury depends on chronic exposure to it (as in some industrial processes where the victim is in daily contact with the toxin); the danger of methyl mercury poisoning resides in the total life dose.

In addition to methyl mercury produced by methylating bacteria from inorganic mercury, effluent from the Minamata factory included methyl mercury as well as the inorganic form. In the 1950s, this factory manufactured acetaldehyde by a process using mercuric sulphate as a catalyst. Typically 300–1000 g of mercury is lost for each tonne of acetaldehyde produced by this method and about 5 per cent of it is in the form of methyl mercury. Mercuric chloride is used as a catalyst in the manufacture of vinyl chloride and about 1 g of methyl mercury per tonne of product is lost in the washing process. Investigations carried out in Minamata Bay in 1959 revealed contamination of sediment as high as 200 p.p.m. mercury near the factory outfall, declining to 12 p.p.m. at some distance. Plankton contained 5 p.p.m., bivalves from intertidal areas contained 10–39 p.p.m. (dry weight), and fish 10–55 p.p.m. (dry weight); most of the mercury in the fish was methylated.

Public health standards

Following the Minamata disaster there was a greater appreciation of the risk of mercury poisoning from eating contaminated seafood, and several countries introduced legal standards for mercury in food offered for sale. The World Health Organization (WHO) recommended a maximum tolerable consumption of mercury in food of 0.2 mg of methyl mercury or 0.3 mg of total mercury per week. Any standard must take into account how much fish is eaten as well as the concentration of mercury in it. In Japan, for example, a far wider range and quantity of seafood is eaten than in Britain where the average per capita consumption of fish is only 20 g per day.

Standards then adopted for the maximum permitted levels of mercury in seafood were 0.5 μg g^{-1} (p.p.m.) in the United States and Canada; 0.7 μg g^{-1} in Italy; 1.0 μg g^{-1} in Germany, Japan, Sweden, and Switzerland; and 1.5 μg g^{-1} in Norway. No standard was introduced in Britain at that time because of the low consumption of fish which generally contains no more than 0.3 μg g^{-1}. Since 1986, a uniform standard of 0.3 μg g^{-1} has been adopted in the EC.

Sweden initially intended to introduce a limit of 0.5 μg g^{-1} but discovered that much of the fish in the Baltic and inland lakes contained more than this, and to have adopted such a stringent standard would have excluded much otherwise wholesome fish from the market. Instead, the limit of 1.0 μg g^{-1} was set, but the public was advised not to eat more than two fish meals per week.

In 1970, it was discovered that samples of canned tuna on supermarket shelves in the United States contained more than 0.5 μg g^{-1} mercury. This led to a temporary withdrawal of tuna until it was discovered that only a small proportion of cans (presumably of older, larger fish) had such a high mercury content.

CADMIUM

Sources and inputs

Cadmium is widely distributed in the earth's crust, but is particularly associated with zinc and it is produced commercially only as a by-product of zinc smelting. Cadmium has been used in quantity since about 1950 and total world production is about 15 000–18 000 t yr^{-1}. The principal uses of cadmium are as stabilizers and pigments in plastics and in electroplating, but substantial amounts are used in solders and other alloys and in batteries.

Less than 10 per cent of the cadmium used in these primary products is recycled and the rest must be assumed to be released to the environment. Although they cannot be quantified, other inputs to the environment are from a variety of often diffuse sources:

(1) fumes, dust, and waste water from lead

and zinc mining and refining, as well as from cadmium production;

(2) rinsing water from electroplating with 100–500 p.p.m. cadmium;

(3) the iron, steel, and non-ferrous metal industries produce dust, fumes, waste water, and sludge containing cadmium;

(4) zinc used in galvanized coatings of metals contains about 0.2 per cent cadmium as an impurity; it is estimated that all this cadmium is lost to the environment through corrosion within four to twelve years;

(5) wear of automobile tyres which contain 20–90 p.p.m. cadmium as an impurity in the zinc oxide used as a curing accelerator;

(6) phosphate rocks may contain 100 p.p.m. cadmium, and phosphate fertilizers are a source of cadmium in the environment;

(7) coal contains 0.25–5.0 p.p.m. and heating oils on average 0.3 p.p.m. cadmium, an unknown amount of which is discharged to the atmosphere;

(8) sewage sludge contains up to 30 p.p.m. cadmium.

The total input of cadmium to the world's oceans is estimated to be nearly 8000 t yr^{-1}, about half of it the result of man's activities, the rest natural. Rivers and atmospheric inputs are equally important. About 2900 t yr^{-1} of the cadmium is deposited in bottom sediments (most on the continental shelf) but it is difficult to account for the fate of the rest. Known fates of cadmium in the sea do not balance the marine budget and the cadmium content of the sea may be slowly increasing.

Cadmium in marine organisms

Cadmium is not an essential element for any organism athough, for unknown reasons, it enhances phytoplankton photosynthesis and growth at concentrations up to 100 p.p.m. Because of its associations with phosphates, cadmium is assumed to be taken up by phytoplankton but, except in two instances, it does not appear to accumulate in the food chain. The euphausiid *Meganyctiphanes norvegica*, for example, feeding on phytoplankton containing 2.1 p.p.m. (dry weight) of cadmium, produces faecal pellets containing 9.6 p.p.m. (dry weight), but has a whole-body concentration of only 0.7 p.p.m. (dry weight). At higher levels in the food web, fishes and sea mammals have low concentrations of cadmium, at most a few p.p.m., stored chiefly in the kidney, and they are able to detoxify it by the production of metallothionein.

Zooplankton in the surface layers of the ocean evidently have high cadmium levels because petrels and the hemipteran sea skater *Halobates*, both of which feed on surface zooplankton far from sources of contamination, tend to have high concentrations of cadmium. Petrels may contain 49 p.p.m. (dry weight) in the liver and 240 p.p.m. (dry weight) in the kidney; *Halobates* normally contains 33 p.p.m., but may contain as much as 330 p.p.m. (dry weight) cadmium.

Molluscs accumulate large concentrations of cadmium. This is particularly so in the bivalve Pectinidae: *Pecten novae-zeelandiae* has been found with 2000 p.p.m. (dry weight) in the liver, but 1900 p.p.m. (dry weight) has been found in the oceanic squid *Symplectoteuthis oualaniensis*, and oysters, limpets (*Patella vulgata*), and dog whelk (*Nucella lapillus*) also acquire high concentrations of cadmium. *Nucella* shows clear evidence of accumulation of cadmium: in the Bristol Channel this dog whelk was found to contain 38 p.p.m. of cadmium, but the barnacles on which it feeds contained only 0.15 p.p.m.

Studies in the cadmium-contaminated Severn estuary and Bristol Channel in southwest Britain, where the input is from natural sources, and in the Sörfjord, a branch of the Hardanger fjord in Norway which receives smelter wastes, have failed to show any ecological effect beyond the contamination of some members of the local fauna.

Itai-itai disease and public health

Cadmium achieved notoriety in the aftermath of the Minamata disaster when it was suspected of being responsible for an out-break of itai-itai disease in a Japanese village on the Jintsu river. This painful disease affected particularly the bones and joints and resulted in 100 deaths. It was attributed to contamination of rice by cadmium from the effluent from a zinc smelter. There is now doubt, however, whether itai-itai is cadmium-related or is more likely to have been associated with malnutrition and vitamin deficiency. High concentrations of cadmium (173 p.p.m.) and zinc (57 600 p.p.m.) in oysters, *Crassostrea gigas*, from the Derwent estuary in Tasmania caused nausea and vomiting in people consuming them, but there is no evidence of people being permanently affected by eating fish or shellfish contaminated with cadmium. With the exception of oysters, scallops, and probably other bivalves from contaminated water, seafood contains no more cadmium than other foods and does not represent a special hazard.

Cadmium and its compounds, along with mercury and some other dangerous metals, are, however, included in the 'black list' of materials which by international agreement may not be discharged or dumped into the sea.

COPPER

Sources and inputs to the sea

The natural input of copper to the marine environment from erosion of mineralized rocks is estimated to be 325 000 t yr^{-1}. Inputs from human activities are localized and vary widely in their nature. About 7.5 million t yr^{-1} of copper is produced for use in electrical equipment, in alloys, as a chemical catalyst, in antifouling paint for ships' hulls, as an algicide, and as a wood preservative. Several of these uses inevitably result in copper being transferred to the environment. Urban sewage contains a substantial amount of copper and this is reflected in enhanced concentrations in sediments at sludge dumping grounds (see Figs 2.9(*c, d*) and 2.12(*c*)). The municipal waste from Los Angeles is estimated to contribute 510 t of copper to the sea annually. Run-off from mine tailings in tin and copper mining areas results in high copper and zinc concentrations in the river Fal in Cornwall and Rio Tinto in Spain. Antifouling paint releases all its copper to the sea. This is not a negligible amount: some antifouling paints contain 500 g l^{-1} of copper and it is estimated that 180 t yr^{-1} of copper entered the coastal waters of California between Santa Barbara and San Diego from this source alone, until copper was replaced by tin compounds in antifouling paints.

Copper dissolved in seawater is chiefly in the form of $CuCO_3$ or, in water of reduced salinity, as $CuOH^+$. It also forms complexes with organic molecules. However, copper is one of the metals readily removed from solution by adsorption to particles and it is estimated that 83 per cent of copper in the sea is in this form.

Copper in marine organisms

Copper is an essential element for animals and the highest concentrations are found in decapod crustaceans, gastropods, and cephalopods, in which the respiratory pigment haemocyanin contains copper. Excess copper is usually stored in the liver; 4800 p.p.m. copper has been detected in the liver of *Octopus vulgaris* and 2000 p.p.m. in the hepatopancreas of a lobster, *Homarus gammarus*. Oysters may acquire very high concentrations of copper, stored mostly in the wandering leucocytes, and these blood cells may contain 20 000 p.p.m. copper and 60 000 p.p.m. zinc.

Although plankton, fish, and shellfish from areas known to be contaminated contain higher concentrations of copper than those from uncontaminated areas, copper does not generally accumulate in food chains. A predatory fish, the marlin (*Makaira indica*) at the top of a food chain accumulates mercury (see p. 69) but has low concentrations of copper: 0.3–1.2 p.p.m. (average 0.4 p.p.m., wet weight) in muscles and 0.5–22.0 p.p.m. (average 4.6 p.p.m., wet weight) in the liver.

Despite the existence of a number of detoxifying and storage systems for copper, it is the most toxic metal, after mercury and silver, to a wide spectrum of marine life, hence its value in antifouling preparations. Its damaging effect is illustrated by the case of a quantity of copper sulphate illegally dumped on the Dutch coast near Noordwijk in 1965. A body of water containing more than 0.3 p.p.m. copper moved slowly up the coast, causing the death of plankton, fish, and shellfish (Fig. 5.4). There were fears that it would reach the extensive shellfish beds in the Waddensee where the economic losses would have been very great. Fortunately, strong winds pro-

duced the necessary mixing and dilution, and this catastrophe was averted.

The very long-standing contamination of the estuary of the river Fal in south Cornwall, southwest England, has been investigated in detail, with interesting results. One branch of the estuary, Restronguet Creek (Fig. 5.5), is fed by the Carnon river which drains a formerly very productive tin mining area. Alluvial tin has been recovered from Restronguet Creek for several thousand years since the Bronze Age, and the river has received wastes from deep mining for several

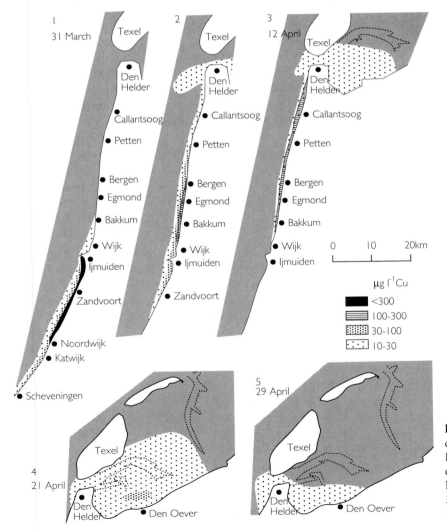

Fig. 5.4 Movement of copper sulphate on the Dutch coast after being dumped on the beach at Nordwijk, from 31 March to 29 April 1965.

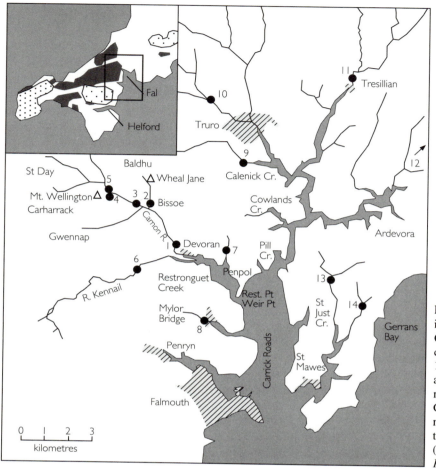

Fig. 5.5 River Fal and its tributaries. The Carnon river drains the chief tin mining area: 1–14 are sampling areas, △ indicates tin mines. Inset: west Cornwall, showing metalliferous areas and the study area. (*Cambridge University Press*.)

hundred years. In the nineteenth century, arsenic refining and lead and tin smelting were carried out in the Carnon valley, adding to the contamination of Restronguet Creek. Although mining activity is now reduced, drainage from ancient mine tailings continues to carry a heavy load of metals into the creek. Levels of contamination of sediments are: zinc 2700 p.p.m., copper 2148 p.p.m., and arsenic 1732 p.p.m. These very high levels of contamination are about two orders of magnitude greater than for uncontaminated estuaries.

Despite this great contamination, the fauna and flora of Restronguet Creek is unexpectedly normal. Only bivalve molluscs, with the exception of *Scrobicularia plana*, and the estuarine gastropod *Hydrobia ulvae* are conspicuously absent. A variety of factors explains the presence of organisms in the water and sediment which might have been expected to be lethal.

Concentrations of copper in the polychaete *Nereis diversicolor* are closely related to those in the sediment and worms containing more than 1000 p.p.m. have been found in the creek. The copper is stored in the epidermal cells. Such levels would be lethal to *N. diversicolor* from uncontaminated estuaries (see Fig. 4.2) and it appears that a copper-resistant strain of this species exists in the Fal. The alga *Fucus vesiculosus* in upper reaches of the creek accumulates several thousand p.p.m. of copper and zinc (Fig. 5.6) and may also be a

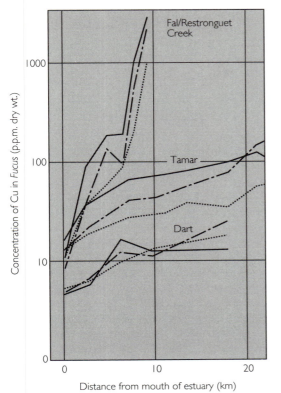

Fig. 5.6 Copper in brown seaweeds. *Fucus*, in three rivers in southwest England: Restronguet Creek receives mine waste, the Tamar passes through old tin mining areas, and the Dart is only slightly influenced by previous mining. ⋯⋯ growing tips; ——— thallus; – · – · stipe.

tolerant strain. Resistant strains are also known to occur in the fouling alga *Ectocarpus siliculosus* which succeeds in growing on ships treated with copper-containing paint.

The polychaetes *Nephtys hombergi* and *Glycera convoluta* and the crab *Carcinus maenas* also show increased tolerance to copper but are mobile species with distributed larvae, and they colonize the creek from less contaminated areas. There may be some selection of tolerant individuals, but copper tolerance may also be induced in these species through exposure to sub-lethal concentrations as they migrate into the creek.

Larval bivalves are very sensitive to copper

and also to zinc, and it is not surprising that they are unable to survive in the high concentrations prevailing in the creek. *Scrobicularia*, which does occur there, is confined to the highest tidal levels where it is exposed to much-diluted river water. Breeding of this species is in July and August, and in dry autumn conditions with a low river input, juveniles have a chance to establish themselves.

Flounders, *Platichthys flesus*, feeding in the creek, largely on *Nereis diversicolor*, appear to be able to limit the assimilation of copper in the gut and do not acquire an increased body burden of the metal, although there is some evidence of liver damage in flounders exposed to high levels of metals.

Redshanks, *Tringa totanus*, wintering in the area, feed almost exclusively on *Nereis diversicolor* and the daily intake of worms by these birds is about equal to their body weight. They receive a high concentration of metals in their diet, but it is not known if they are affected by it.

Public health

Bivalves growing in contaminated water have the capacity to accumulate copper: the concentration factor for oysters is 7500 and they may accumulate so much copper that the flesh becomes green. This commonly happens to the oyster *Ostrea edulis* in the river Fal and they have to be relaid in uncontaminated water for a year before they can be marketed. There is not danger to humans of copper poisoning from seafood—the lethal dose is about 100 mg—but the human taste threshold for copper is low, 5.0–7.5 p.p.m., and the taste is repulsive. There is an ample safety margin and copper in the sea is not regarded as a health hazard.

LEAD

Sources and inputs

The total world production of lead is about 43 million t yr^{-1}. Much lead in metallic form, in battery casings and plates, in sheet and pipes, and so on, is recovered and recycled, but most lead used in compound form is lost to the environment. Nearly 10 per cent of the world production of lead is used as petrol additives such as lead tetraethyl and it is lost, largely to the atmosphere. Globally, inputs to the atmosphere resulting from human activities, 450 000 t yr^{-1}, dwarf natural inputs of 25 000 t yr^{-1}.

Lead aerosols are carried to earth in rain and snow and are widely scattered. Figure 5.7 shows the lead content of the annual ice layers in Greenland from this source. The start of the industrial revolution marks the first major increase in the deposition rate; the growth in the number of automobiles using leaded petrol since 1940 is reflected in a second marked acceleration in deposition rates. The lead content of a peat profile in Derbyshire (Fig. 5.8) shows a similar increase.

Local high concentrations of lead may be caused by special circumstances. Sea bass taken on the Californian coast near Los Angeles, with its very high density of automobiles, contain 22 p.p.m. (wet weight) of lead in their livers. Corresponding figures for fish caught 300 miles offshore have 10 p.p.m., and off the Peruvian coast 9 p.p.m.

Sewage sludge dumping grounds may be expected to contain high lead concentrations. Glasgow sewage sludge contains 771 p.p.m.

lead and sediments in the dumping area 200–320 p.p.m., whereas the general background in the Firth of Clyde is 48–139 p.p.m. (Fig. 2.9(b)), and in an area far from sources of contamination such as Trinidad, is only 22 p.p.m.

Impact on marine organisms

Compared with other metals, lead in the sea is not particularly toxic and, at concentrations up to 0.8 p.p.m., lead nitrate even enhances the growth of the diatom *Phaeodactylum* (Fig. 5.9), presumably through the nutrient effect of the nitrate. Sub-lethal effects of low concentrations include a depression of the growth of *Cristigera* (a ciliate protozoan) by 8.5 per cent at 0.15 p.p.m. and 11.8 per cent at 0.3 p.p.m. (Table 4.3, p. 58); the growth of the crustacean *Artemia* is significantly reduced at

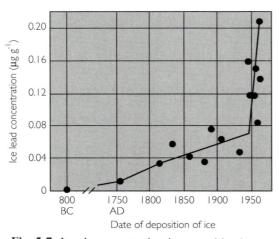

Fig. 5.7 Lead concentration in annual ice layers in Greenland. (*Pergamon Press.*)

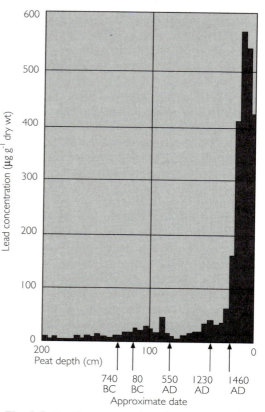

Fig. 5.8 Lead concentrations in a peat profile in Derbyshire. (© *1973, Macmillan Journals Ltd.*)

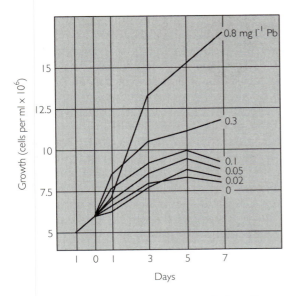

Fig. 5.9 Growth of a culture of the diatom *Phaeodactylum* with lead nitrate in various concentrations added after the first day.

5–10 p.p.m.; and the mortality rate of the mussel *Mytilus edulis* is increased by prolonged exposure to 10 p.p.m. or less of lead salts.

Nevertheless, high concentrations of lead can be accumulated by some animals without apparent harm. The limpet *Acmaea digitalis* may contain about 100 p.p.m. on the Californian coast near San Francisco where evidently the contamination is from aerial fall-out. In the river Gannel in north Cornwall, where there is a natural input from lead deposits, estuarine sediments contain 2175 p.p.m. lead and the bivalve *Scrobicularia plana* living in them, contains 991 p.p.m. In the contaminated Sörfjord in Norway, seaweeds and animals contain high levels, the content in mussels rising to 3000 p.p.m.

Mytilus has a detoxifying mechanism for lead and stores large quantities in the form of granules in the digestive gland. Because of their tendency to accumulate metals, mussels are regularly monitored in western European waters by the International Council for the Exploration of the Sea. Generally, the concentration of lead is below 1 p.p.m. (wet weight), though occasional samples with up to

4 p.p.m. have been recorded on the Swedish west coast. Lead concentrations in mussels have declined in recent years in the North Sea and Baltic, which probably reflects a reduction in the use of leaded petrol.

Fish contain little lead and the content in commercial species in the North Sea generally ranges from 0.05–0.15 p.p.m. (wet weight). The livers of fish-eating seals and porpoises contain no detectable lead. Evidently, it does not present a hazard through bioaccumulation.

A rare instance of lead poisoning occurred among shore birds wintering on the Mersey estuary: 2400 birds, mainly dunlin (*Calidris alpina*), died in 1979 and a smaller number the following year. The dead birds contained more than 10 p.p.m. (dry weight) of lead in their liver, 30–70 per cent of which was in the form of trialkyl lead. Further investigations showed that birds with 0.5 p.p.m. trialkyl lead in the liver were at risk and some birds showed neuromuscular disorders symptomatic of lead poisoning. These waders acquired the lead from their food, which was predominantly the bivalve *Macoma baltica* containing 1 p.p.m. lead, and the polychaete worm *Nereis diversicolor* containing 0.2 p.p.m. It is thought that a discharge of organic lead in factory effluent was responsible for contaminating the benthic fauna of the estuary.

Public health

Although lead is held responsible for serious damage to health on land, such contamination of the sea and marine products as occurs does not appear to be a matter for concern.

IRON

Iron is not usually a significant contaminant of the sea, but it has come to prominence over sea dumping of red mud from the extraction of alumina from bauxite, and acid iron waste from the production of titanium dioxide; both wastes contain a large amount of iron. It is unfortunate that titanium dioxide manufacture should create environmental problems because it was introduced as a white pigment in paints to allow lead pigments to be phased out, so reducing human exposure to lead. Very large quantities of these wastes are produced: western Europe produces about 7.5 m t yr^{-1} of acid iron waste, of which 5.6 m t are discharged to sea, either directly by dumping or indirectly through pipelines and rivers. Eastern Europe discharges about 2 m t yr^{-1} to the Baltic and Black Seas and inland waters.

Because acid iron waste is discharged in such quantity, great concern was expressed about its possible environmental impact by the fishing industry and by conservation bodies. In response, the US Environmental Protection Agency recommended that the waste should no longer be dumped in New York Bight, but 90 miles further out to sea. Germany and Holland curtailed the dumping of their wastes into the North Sea.

Despite these precautionary measures, it has proved difficult to identify the environmental impact of acid iron waste discharges. The iron salts are precipitated as hydrated oxides of iron which drift around as flakes or particles before settling. A heavy deposit of iron oxides, like rust, forms on solid objects, including the shells and carapaces of animals. Particles adhere to fish eggs and larvae when they may clog delicate feeding structures. In one case, hydrated ferric oxide caused loss of weight and increased mortality in mussels, *Mytilus edulis*, apparently through the increased production of pseudofaeces and loss of organic matter through the secretion of additional mucus. The oxides also appear to precipitate on the gills of fish, and other metals coadsorbed with the iron are readily taken up.

It is impossible to reproduce in the laboratory the behaviour of iron oxides in the sea, however, and studies at dump sites yielded uncertain evidence about the impact of the iron in the wastes. A study of the dumping grounds in New York Bight in the early 1970s, after some 50 m t of titanium dioxide waste had been discharged over the previous 20 years, could detect no particular effects attributable to the iron content. On the other hand, this dump site had received a wide variety of toxic wastes for many years, it had an impoverished fauna, and only hardy species remained. Studies of plankton, benthos, and fish at a German dumping ground in the North Sea also failed to show any change, although iron concentrations in the area had increased. An attempt was made to relate the incidence of epidermal tumours and fin rot in fish to acid iron waste dumping in the area, but these conditions may arise in any area of waste disposal and appear to be a response to general stress and not related to a particular contaminant.

Attention has since focused on the acid component of the waste and an EC directive calls for a reduction in acid (sulphate) discharges by the titanium dioxide industry by 1992. However, the acid is rapidly neutralized in the sea and has, at most, only a very localized effect. There is now evidence from studies made in several parts of the world that trace metal residues, rather than the acid, are responsible for some biological disturbance. Increased concentrations of lead, vanadium, chromium, zinc, and iron (in decreasing order of importance) in bottom sediments around the outfall are correlated with a reduction of diversity and population size of the meiofauna (Fig. 5.10). Iron hydroxide in solution causes sperm agglutination, which results in reduced fertilization, and this could obviously have serious implications for organisms that practise external fertilization; however, there is no evidence that acid iron waste discharges have affected local populations for this reason.

Contrary to previous fears, acid iron wastes appear to cause only a small and localized perturbation around the point of discharge, and iron is a less significant contaminant than other trace metals included in the waste.

TIN

Antifouling paints containing organic tin compounds, tributyltin oxide (TBT) and

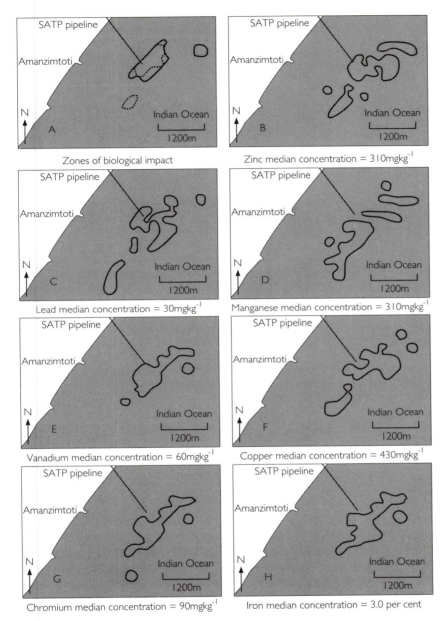

Zones of biological impact

Zinc median concentration = 310mgkg^{-1}

Lead median concentration = 30mgkg^{-1}

Manganese median concentration = 310mgkg^{-1}

Vanadium median concentration = 60mgkg^{-1}

Copper median concentration = 430mgkg^{-1}

Chromium median concentration = 90mgkg^{-1}

Iron median concentration = 3.0 per cent

Fig. 5.10 Survey area off the Natal coast, near Durban, including a titanium dioxide waste outfall, showing (*A*) the zone of modified community structure (continuous line) and of reduced population density (broken line) of the meiofauna, and concentrations in bottom sediments of (*B*) zinc, (*C*) lead, (*D*) manganese, (*E*) vanadium, (*F*) copper, (*G*) chromium, and (*H*) iron.

tributyltin fluoride, are much more effective and long-lasting than copper-based anti-fouling paints. They came into use in all classes of shipping in the 1970s and sub-sequently were also commonly used to treat net enclosures of mariculture installations and wooden lobster pots.

TBT is extremely toxic and is lethal to a variety of planktonic organisms, including mollusc larvae which are 10–100 times more sensitive than the adults, at concentrations of only 1.0 μg l^{-1}. Such concentrations are commonly found in waters in and around yacht marinas and have been associated with recruitment failures of scallop and oysters in the affected areas.

The sub-lethal effect of low concentrations of TBT are of even greater consequence to commercial shellfisheries. Concentrations of TBT down to 0.01 μg l^{-1} cause reduced growth and other sub-lethal effects in very young Pacific oysters *Crassostrea gigas*. More seriously, TBT causes gross thickening of the shells of oysters, greatly reducing the size of the animal inside, and rendering them unmarketable.

A curious and equally damaging sub-lethal effect of TBT has been discovered in intertidal, neogastropod dog whelks. A penis and sperm duct are developed in females, a condition known as 'imposex', evidently as a result of TBT triggering a change in the hormonal system. The oviduct and female genital opening are eventually overgrown by the developing penis and are blocked so that the egg capsules cannot be laid. Since dog whelks do not have a planktonic or other distributive stage in their life cycle, affected populations gradually decline and may be eliminated altogether because of this failure of reproduction. Most of the detailed studies have been made of the dog whelk *Nucella lapillus*, which has been seriously affected on northwest European rocky coasts exposed to inshore shipping traffic. It has now been found that this is a worldwide problem and neogastropods (*Nucella*, *Thais*, *Cronia*, *Ilyanassa*, *Drupella*, *Naquetia*, and so on) are similarly affected in the northwestern United States and Canada, Singapore, Malaysia,

Indonesia, and New Zealand.

It is not known if TBT has damaging effects on other marine organisms, but that cannot be ruled out.

In mariculture installations where TBT is used to prevent fouling of nets and enclosures, organotin concentrations of 0.28–0.90 μg g^{-1} (wet weight) have been detected in the flesh of chinook salmon reared in them. This organotin is not destroyed by cooking.

Because of the damage to shellfisheries, environmental damage, and the possible threat to human health, the use of TBT in mariculture installations and on small boats less than 25 m in length has been banned in several countries, including France, Ireland, the United Kingdom, and the United States. An EC directive with the same prohibitions came into effect in 1991. For the present, the use of TBT in antifouling paint used on ocean-going vessels has been permitted. This is on the grounds that such large vessels do not congregate in inshore waters as small boats do, and the use of TBT antifouling paints on them is unlikely to result in a damaging concentration of TBT near shellfish beds or rocky coasts.

TBT degrades in seawater to non-toxic compounds within a few weeks and the problems it has caused are unlikely to have a lasting effect. On the northeast coast of England, populations of *Nucella lapillus* suffered a high incidence of imposex and had declined as a result. Four years after the ban on the use of TBT, populations showed a substantial recovery.

6

HALOGENATED HYDROCARBONS

Hydrocarbons containing fluorine, chlorine, bromine, or iodine (the halogens) differ from petroleum hydrocarbons (Chapter 3) because most of them are not readily degraded by chemical oxidation or bacterial action. Like metals (Chapter 5), they are essentially permanent additions to the environment but, unlike metals, most of these compounds are man-made and do not occur naturally. The great majority contain chlorine and are known collectively also as chlorinated hydrocarbons or organochlorine compounds.

LOW-MOLECULAR-WEIGHT COMPOUNDS

Halogenated hydrocarbons embrace a very wide range of compounds. Low-molecular-weight hydrocarbons, particularly methane, are synthesized by marine algae and possibly by a few invertebrates, and may contain chlorine, bromine, or occasionally iodine. Elevated concentrations of these substances in seawater may therefore come from natural sources and not be the result of human activity.

Low-molecular-weight, volatile, halogenated hydrocarbons are, however, manufactured in large quantities and almost all this production is eventually lost to the environment. Those produced in greatest quantity, totalling some 30 m t yr^{-1}, are the industrial solvents dichlorethane (CH_3CHCl_2) and vinyl chloride ($H_2C=CHCl$). Other solvents with a world production approaching 1 m t yr^{-1} each are carbon tetrachloride (CCl_4) and perchlorethylene ($Cl_2C=CCl_2$) used in dry cleaning, and trichlorethane (CH_2Cl_3) and trichlorethylene ($ClHC=CCl_2$). Chlorofluorocarbons or freons (CCl_3F and CCl_2F_2) are used as coolants, aerosol propellants, and in foamed plastics (Table 6.1). World production of freons in 1986 amounted to over 1 m t (Table 6.2).

Some of these compounds have become very widely distributed in the atmosphere and can even be detected in the air of Antarctica, far from any possible source. They occur in detectable quantities in seawater close to centres of industrialization. There is concern about the effects of freons on the ozone layer in the upper atmosphere and strenuous

Table 6.1 European Community sale of freons (thousand t)

Use	1976	1977	1978	1979	1980
Aerosols	176.9	162.6	150.4	136.6	126.5
Refrigeration	21.8	20.3	20.4	20.3	21.2
Foams	42.2	45.3	54.5	55.8	61.9
Other	4.2	4.9	6.1	6.9	7.4
Total	245.1	233.1	231.4	219.6	217.0

Table 6.2 World production of freons in 1986 (thousand t)

Country	Production
European community	438
United States	295
Japan	125
USSR	100+
China	20
Others	45
Total	1038

efforts are being made to reduce production, but low-molecular-weight hydrocarbons in the sea are not regarded as a particularly serious threat. They appear not to accumulate in marine organisms.

PESTICIDES AND PCBS

Attention has been focused particularly on the higher-molecular-weight chlorinated hydrocarbons which do accumulate in animal fat tissues. These include pesticides and polychlorinated biphyenyls (PCBs) (Fig. 6.1).

DDT and related compounds

Although DDT (dichloro-diphenyl-trichloro-ethane) was known in the last century, it was not introduced as an insecticide until 1939. In many ways it is the ideal pesticide:

(1) it is extremely toxic to insects, but very much less toxic to other animals;

(2) it is very persistent, with a half-life in the soil of about 10 years;

(3) it continues to exert its insecticidal properties for a very long time;

(4) it is relatively cheap.

Few other pesticides share the desirable properties of being reasonably specific to the target organisms, of remaining effective for a long time so that only a single application may be needed, and of being safe for the humans exposed to it.

The use of DDT prevented epidemics of typhus among refugees and in war-damaged areas in Europe during and immediately after the Second World War. It provides the basis of the very successful programmes of the World Heath Organization to control malaria and other insect-borne diseases in Africa and the Far East. Its primary use, however, particularly in developed countries, was pest control in forestry and in most forms of agriculture.

DDE (dichloro-diphenyl-ethane) is a derivative of DDT, involving the loss of one chlorine atom from the —CCl₃ group in the DDT molecule. It has low toxicity to insects and is not used as a pesticide. Most of the chlorinated hydrocarbon in the sea and 80 per cent of that in marine organisms is in the form of DDE and presumably nearly all of it has been derived from the breakdown of DDT.

DDD   (dichloro-diphenyl-dichloroethane)

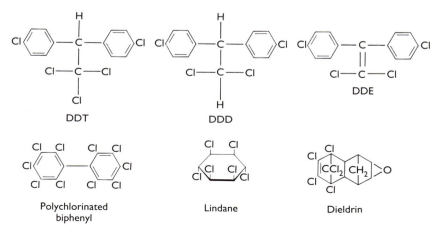

Fig. 6.1 The structure of some chlorinated hydrocarbons.

is another derivative of DDT involving the loss of a chlorine atom from the —CCl_3 group in the DDT molecule. It has some toxicity to insects and is less toxic than DDT to fish. For this reason it is occasionally used as an insecticide in situations where its low toxicity to fish is important. It can be excreted by many organisms and rarely accumulates in them.

Commercial DDT is a mixture of DDT, DDE, and DDD, though with DDT predominating.

DDT has now been almost completely superseded by less persistent insecticides in the developed world, but it continues to be extensively used in developing countries because it is effective, safe to handle, and cheap.

'Drins'

This group of interrelated insecticides includes aldrin, dieldrin, endrin, heptachlor, and so on. They are all extremely persistent and the degradation products are also toxic: heptachlor degrades to heptachlor epoxide which is even more toxic than the parent compound; aldrin degrades to dieldrin.

These pesticides were used as seed dressings and in situations where their high toxicity and persistence were required, as in the control of wireworms (*Tipula* larvae) in ploughed-up grassland. Their chief disadvantage is that they are toxic to mammals as well as being persistent and subject to bioaccumulation. They were largely withdrawn during the 1970s although they continued to be used to moth-proof textiles, carpets, and fleeces, and for some minor agricultural purposes, until the late 1980s.

Although now used only in very small quantity because of their persistence, 'drins' are widespread in the environment and continue to leach out of agricultural land into water courses and the sea.

Lindane (γ-HCH)

Lindane (gamma-hexachlorocyclohexane, or γ-HCH) was previously, and sometimes still is, incorrectly known as benzene hexachloride (BHC) and this name may even be erroneously applied to hexachlorobenzene (see below).

Lindane came into use at about the same time as DDT and acts as a contact poison to insects. It is volatile and stable at high temperatures and so can be used as a smoke to fumigate crops in, for example, orchards. It is also used as a seed dressing and in the treatment of timber. It tends to accumulate in food chains and some attempt has been made to restrict its use; γ-HCH levels in cod livers in the southern North Sea fell from about 300 μg kg^{-1} in 1977 to about 50 μg kg^{-1} in 1987. However, it is still widely employed in agriculture and horticulture, particularly in India and China.

Hexachlorobenzene (HCB)

This product was formerly widely used as a soil fumigant and as a seed dressing for grain or as a fumigant in grain storage against fungal attacks. It is also used in wood preservatives. It has largely been replaced as a soil fumigant, but continues to be used in agriculture for other purposes. It is used as a fluxing agent in aluminium smelting and also occurs as a by-product in the manufacture of carbon tetrachloride, pentachlorophenol, and vinyl chloride monomer. There are therefore numerous routes by which it may reach the sea. It is relatively insoluble in water and most of the HCB in the sea is attached to sediment particles. It is highly persistent.

Toxaphene

Toxaphene is the trade name for polychlorinated camphenes and was introduced as an insecticide in the mid-1940s. Its method of synthesis results in the production of a complex mixture of at least 670 camphene compounds with six to ten chlorine atoms per molecule and a total chlorine content of about 68 per cent by weight. It is used as an insecticide on cotton and vegetable crops, and in livestock dips. It is acutely toxic to fish and is occasionally used as a piscicide in freshwater lakes. Since its introduction it has been the most heavily used pesticide in the United

States, with a total production not far short of that of DDT at its peak. Its use is now coming under regulation.

Polychlorinated biphenyls (PCBs)

PCBs have been in use since the early 1930s. They are not pesticides, but are used in electrical equipment, in the manufacture of paints, plastics, adhesives, coating compounds, and pressure-sensitive copying paper. Because they are chemically very stable, they resist chemical attack and act as flame retardants. They have been used in transmission fluids in fluid-drive systems, and as dielectrics in transformers and large capacitors.

The number of chlorine atoms per molecule varies from one to ten, but commercial PCBs are produced to physical, not chemical, specifications and contain a mixture of isomers. The product may contain 20 per cent by weight of chlorine and average about one chlorine atom per molecule, or up to 60 per cent chlorine with different percentages of biphenyls containing three to six atoms of chlorine per molecule.

Following concern about the environmental accumulation of chlorinated hydrocarbon pesticides, a number of steps were taken to reduce the use of PCBs. In 1970, Monsanto, sole manufacturer of PCBs in the United States, voluntarily reduced production from 33 000 t in 1970 to 18 000 t in 1971. Since then, a variety of measures have been adopted to limit the use of PCBs to functions where recovery and recycling is possible: dielectrics in transformers and large capacitors and, in some countries, in hydraulic systems. In 1985, the Netherlands stopped using PCBs altogether. Up to the time that the use of PCBs came under restriction, however, total world production had amounted to more than 1 m t, most of which has been dispersed in the natural environment.

Dioxins and furans

There are 75 different isomers of chlorinated dioxins containing one to eight chlorine atoms. The isomer TCDD (2,3,7,8 tetra-chlorodibenzodioxin) has caused the greatest concern. Chlorinated dibenzofurans are structurally similar to dioxins, but there are 135 isomers.

Possible sources of dioxins and furans in the environment include:

(1) manufacture and use of chlorophenols especially in the wood processing and treatment industries;

(2) combustion of organic material, including fossil fuels, wood, municipal wastes, and so on;

(3) manufacture and use of the herbicides 2,4,5-T and 2,4-D, which are a significant source of di-, tri-, and tetra-CDDs.

Dioxins are physically and biologically stable, and tend to be increasingly stable with increasing halogen content. They are insoluble in water but soluble in organic solvents, fats, and oil.

OCEAN INCINERATION

Because of their persistence and toxicity, it is difficult to dispose of chlorinated organic compounds safely. The principle waste is EDC-tar, a liquid tar containing 60–70 per cent chlorine, which is a residue from vinyl chloride manufacture. Between 1970 and 1985, 1.75 m t of EDC-tar were produced in western Europe alone. In addition to EDC-tar, PCBs, and certain herbicides and organo-chlorine insecticides that have been withdrawn from use have to be disposed of.

If such wastes are dumped on land, it should be in a situation where there is no risk that they will leach out and contaminate ground or surface water. There are numerous unsatisfactory landfill sites in many countries where this precaution has not been observed. Exhausted salt mines are ideal and are used in Germany and England, but they do not have the capacity to receive all the wastes for disposal. Until the end of the 1960s, a proportion of these wastes was dumped at sea, but this practice has been abandoned in favour of incineration. Providing that the temperature of combustion is maintained above 1000 °C, there is excess oxygen, and the waste remains an adequate time in the burn zone, then PCBs, dioxins, and so on are destroyed. The principal combustion products are carbon dioxide and hydrochloric acid, but small quantities of heavy metals may be included in the exhaust gases, depending on the nature of the wastes.

Specialized incinerators on land, designed to burn these wastes, have scrubbers to remove the hydrochloric acid from the emissions. But incinerators, particularly those dealing with toxic wastes, are never popular with the surrounding population who fear the consequences of an accidental release of dangerous material.

Partly in response to the difficulty of gaining acceptance of incinerators on land, and partly because it was cheaper, incinerator ships have been developed to burn highly chlorinated wastes at sea. Trial burns have been made in the Gulf of Mexico, the Pacific, and Australia, but ocean incineration has been in regular use only in the North Sea since 1969. During most of the 1980s, 90 000–100 000 t of chlorinated wastes, two-thirds of them from Germany, the rest from a number of other European states, were incinerated each year at a designated site 70 nautical miles from land in the Dutch sector of the North Sea. Incineration was at more than 1200 °C and achieved 99.9999 per cent destruction of PCBs. The practice was regulated and closely monitored by the Dutch authorities.

Exhaust gases from ocean incineration are not scrubbed; hydrochloric acid, the main product, is rapidly absorbed by the sea which has an enormous buffering capacity. The metal input is tiny compared with the input of metals from the atmosphere and rivers (see Table 5.1). It has been suggested that traces of dioxins might be produced in the heated exhaust gas plume, but this has not been established. No environmental damage has been detected from this practice but there is some evidence of an accumulation of organochlorine compounds in bottom sediments in the burn zone.

Despite the advantages of incinerating highly chlorinated wastes at sea, the practice did not receive acceptance outside Europe and it has now been abandoned altogether.

INPUTS TO THE MARINE ENVIRONMENT

Aerial transport is an important route by which chlorinated hydrocarbon pesticides reach the sea and they can be detected in the atmosphere even at remote oceanic sites (Table 6.3).

Table 6.3 Average concentration (in ng m^3) of some chlorinated hydrocarbons in the atmosphere at marine sites

	PCB	Total DDT	Total HCH
South Carolina	0.25	0.036	—
North Atlantic	0.69	0.006	0.39
Enewetak	0.05	< 0.006	0.26
Samoa	0.012	0.0015	0.032

Low-molecular-weight chlorinated hydrocarbons are volatile. Both HCH and HCB are rapidly lost to the atmosphere in the presence of water vapour, and all organochlorine pesticides volatilize and, particularly in the tropics where they are still used in large quantities, climatic conditions favour their release to the atmosphere. In a recent study in south India, 99.6 per cent of the HCH applied to rice paddies was found to be lost to the air. In Nigeria, 98 per cent of the DDT applied to a cow pea crop volatilized in four years.

Further to this, DDT, the 'drins', and toxaphenes also adsorb strongly on to particles and are carried in wind-borne dust.

Other important inputs to the sea are from rivers—over 3 t yr^{-1} of PCBs enter the North Sea by this route, mainly from the Rhine—and from the dumping of dredging spoil or sewage sludge, which is often highly contaminated.

The uses for which pesticides and most PCBs are designed result in them being widely distributed in the natural environment, but the principal source of widespread pesticide contamination is from the agricultural use of these compounds.

1. *Aerial spraying.* In the fruit- and vegetable-growing areas of California where there is heavy use of pesticides, it is estimated that 50 per cent of the pesticide released from crop-spraying aircraft never reaches the ground but forms aerosols and, as such, may travel great distances.

2. *River run-off.* In addition to rain washing pesticide from the plants and soil into rivers and hence to the sea, irrigation water, particularly if it is applied by spraying, has the same effect. Floods carry very large quantities of silt to the sea and, if derived from agricultural land, the silt may carry a considerable quantity of adsorbed pesticide.

3. *Blown dust.* Arid areas may be intensively cultivated with the aid of irrigation, but dry soil with its burden of pesticides is transported in dust storms.

Aerial transport has resulted in the world-wide distribution of organochlorine pesticides. DDT is an entirely man-made substance which does not occur naturally; it has been in use only since 1940. Yet within 20 years, DDT and its residues had pervaded the entire biosphere. Even king penguins in Antarctica, several thousand kilometres from any place where DDT has been used, contain detectable traces of it.

In the manufacture of chlorinated hydrocarbons, a variety of related compounds is usually produced, only one of which is the desired product. During purification, these unwanted hydrocarbons (whose identity is generally unknown) are extracted and were formerly often dumped at sea in considerable quantities.

Locally, chlorinated hydrocarbons appear in industrial effluent in outfalls to the sea. The Montrose Chemical Company in Los Angeles is the world's major manufacturer of DDT. The Los Angeles sewerage system received the effluent from this factory and at peak production in 1971, this resulted in a discharge of 21.6 t yr^{-1} of DDT residues to the sea through the ocean outfall of this sewage system. A survey made in 1972 suggested that 200 t of DDT residues were trapped in the upper 30 cm of bottom sediment over an area of 50 km^2 around the outfall (Fig. 6.2). Discharges from the factory were substantially reduced after 1971 and amounted to only 0.73 t in 1977.

Sewage sludge also may contain appreciable quantities of halogenated hydrocarbons and if dumped at sea provides an additional source of contamination. Sewage sludge from Glasgow in the 1960s made a contribution of 1 t yr^{-1} of PCBs to the Clyde dumping grounds (Fig. 6.3) until brought under control.

FATE IN THE SEA

Chlorinated hydrocarbons are extremely insoluble in water, with a saturation concentration of no more than 1 p.p.b., but they are soluble in fats and adsorb strongly on to particles. Their distribution in the sea is, therefore, far from uniform.

The surface layer of the sea is a film varying from a few μm to 1 mm in thickness. It is extremely difficult to study and is still an area of considerable ignorance, but it is known to contain fatty acids. Because of their lipid solubility, organochlorines may therefore accumulate in it. While the total quantity may not be great, the organochlorine enrichment of the surface film may be of considerable importance to surface-living organisms or to birds, such as petrels, which skim fat off the surface of the sea. Since the sea surface is a site of interchange with the atmosphere, organochlorines may be transferred to the air in aerosol droplets.

Where water masses with different physical or chemical characteristics meet, a front is formed. Oceanic and coastal fronts accumulate floating material, including surface oil. Fronts have a high productivity and attract a wide range of fish, birds, and sea mammals, which therefore receive a diet enriched in organochlorines.

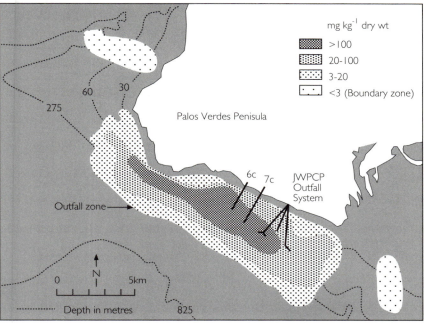

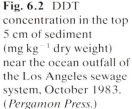

Fig. 6.2 DDT concentration in the top 5 cm of sediment (mg kg⁻¹ dry weight) near the ocean outfall of the Los Angeles sewage system, October 1983. (*Pergamon Press.*)

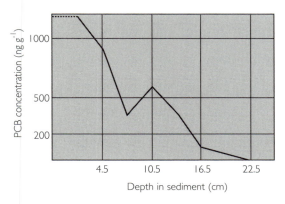

Fig. 6.3 Concentration of PCBs at various depths in the bottom sediment near the Firth of Clyde dumping ground. (*Pergamon Press.*)

A considerable quantity of organochlorines is adsorbed on to particles or on to micro-organisms such as diatoms. This creates problems for analysis because, apart from the difficulty of distinguishing between the various halogenated hydrocarbons, there is the difficulty of deciding what is incorporated in organisms and what is adsorbed on the outside of them; the former may affect them, while the latter cannot but is biologically available to animals that feed on the contaminated organisms.

Chlorinated hydrocarbons adsorbed on to inorganic particles may ultimately be carried to the seabed which then acts as a sink for these compounds. However, suspended or resuspended particles, if they are of a suitable size or density, are commonly ingested by filter-feeding animals and chlorinated hydrocarbons may enter food chains by this route.

Halogenated hydrocarbons, most particularly DDT and its derivatives, now occur in all organisms in all environments, and a considerable quantity in total of the halogenated hydrocarbons in the sea is in the bodies of marine organisms and will continue to circulate within the food webs.

BIOLOGICAL EFFECTS OF HALOGENATED HYDROCARBONS

Problems of analysis

As with petroleum hydrocarbons, halogenated hydrocarbons embrace an enormous variety of related compounds: it may be possible to distinguish 100–150 different organochlorines in a single sample and to identify and measure most of them. Such an abundance of related compounds makes it very difficult to identify which of them is responsible for an observed biological effect. The problem is probably worst for PCBs because the commercial products are mixtures, the exact composition of which varies with the degree of chlorination, the manufacturer, and even with the production batch. Up to 70 PCB compounds may be present in a single sample and the analyst can either estimate total PCBs in the mixture or painstakingly identify and measure individual constituents. The latter approach is very expensive in time and money, so correspondingly fewer samples can be processed, besides which, unless it is known or suspected which of the many compounds is responsible for the biological effect, nothing is gained by the apparently great accuracy of the detailed analysis. The alternative use of cruder measurements of total PCB contamination may be responsible for discrepancies between reports of the biological effects of PCBs. Some other classes of organochlorines, particularly toxaphenes, present the same problem, but much less attention has been paid to them.

A second problem is more fundamental. By 1965, it was possible to distinguish DDT from its metabolites DDE and DDD, and from dieldrin, but it was not until 1966 that PCBs were first identified and a year or two later that their separation from DDT was practised in most analytical laboratories. PCBs had been in use and released to the environment for some thirty years, and they were probably present in all samples. Since PCBs can interfere with the detection and measurement of other organochlorines, many, perhaps most, early records of DDT contamination are incorrect. As late as 1980, toxaphenes, which are also widespread contaminants, were not reliably detected in analyses of chlorinated hydrocarbons. Until recently, it was not possible to identify dioxins and furans reliably. Many chlorinated organic com-

pounds have still not been identified and may well be biologically active. Caution must therefore be exercised in interpreting the effects of halogenated hydrocarbons.

Storage and bioaccumulation

There is evidence that chlorinated hydrocarbons are difficult to excrete and hence they tend to accumulate in the body. Because they are lipid-soluble, they tend to occur in much higher concentrations in fatty tissues than in other tissues. This introduces two risks:

1. In times of poor feeding, animals mobilize and use their fat reserves, so increasing the concentration of chlorinated hydrocarbons circulating in the body, possibly to a dangerous level.

2. As with other bioaccumulating contaminants, there is a strong possibility of transmission through food webs and of biomagnification, as, indeed, has been established in a number of animals.

Because of the tendency of chlorinated hydrocarbons to be sequestered in fatty tissues, care is needed in comparing levels of contamination in different organisms. Different amounts and concentrations are likely to occur in fat tissue, muscles, gonads, and so on, and even if the total body-burden of chlorinated hydrocarbon is known, this has quite different implications for a fat animal from an emaciated animal.

Accumulation rates for organochlorines, measured by residues in the body divided by residues in the food or the environment, vary widely between species. Figure 6.4 shows the average concentrations of PCBs and six organochlorine pesticides in five species living in the contaminated Weser estuary on the North Sea coast of Germany. The actual concentrations acquired relate to the lipid content of the animal and its position in the food chain, and the sole (*Solea solea*), with at least twice as much lipid as the invertebrates, as well as being a carnivore, is the more contaminated as a result. But while the sole, shrimp (*Crangon*), and lugworm (*Arenicola*)

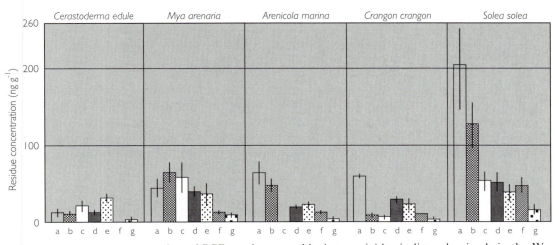

Fig. 6.4 Average concentration of PCBs and organochlorine pesticides in littoral animals in the Weser estuary: (*a*) PCB, (*b*) DDD, (*c*) dieldrin, (*d*) α-HCH, (*e*) γ-HCH, (*f*) DDE, and (*g*) endosulphan. (*Pergamon Press.*)

retain a higher proportion of PCBs than pesticides, the cockle (*Cerastoderma*) acquires relatively little PCB but proportionately much more lindane—which is not preferentially accumulated by the others. *Crangon* is remarkable in accumulating proportionately less DDT and DDE than the other species. Each species, in fact, has its own profile of organochlorine accumulation.

Even within a species there may be considerable variation in the amount of organochlorine that is acquired. Mussels (*Mytilus* spp.) vary greatly in the DDT residues and PCBs they contain (Table 6.4). A number of different factors undoubtedly contribute to this variation; not all are understood but they include age, tidal level at which they live, sex, stage in the annual breeding cycle, as well as the concentration of organochlorines to which the animal is exposed. It is not only molluscs that show such unexplained interspecific variations. In the grey seal (*Halichoerus grypus*), males acquire more DDT residues and PCBs than females, but females more dieldrin than the males.

Distribution of organochlorines within the animal is not uniform. Figure 6.5 shows how DDT is distributed in different tissues of the sole (*Solea solea*) after three days in which

they had been fed 17 μg of DDT labelled with ^{14}C, and after two months in water free from DDT. The brain and, more understandably, the liver both acquire the highest concentrations and eliminate them most slowly.

With so much specific and individual variation it is difficult to generalize, but the highest concentration factors (for DDT) are found in bivalve molluscs, such as oysters and clams, where they may reach 70 000. For crustaceans and fish, accumulation factors range between 100 and 10 000; for sea birds it is 10 or less.

Since organochlorines are eliminated slowly from the body, they may be expected to show biomagnification in food chains. Figure 6.6 shows the average concentrations of organochlorines in land birds, freshwater birds, and sea birds, in relation to the main types of food they eat. Those with most are land birds feeding on other birds (hawks); they live in an environment where pesticides are most immediately available and they are at the top of a food chain.

Biological effects

It is difficult to be precise about the effect of halogenated hydrocarbons on plants and animals. Because of the low solubility of these

Table 6.4 PCB and DDT/DDE concentrations in seawater and *Mytilus*

Marine area and year	Seawater concentration (ng l^{-1})	Concentration in mussel (mg g^{-1})	Concentration factor
PCBs			
France:			
Valras Beach, 1974	0.94	0.25	270 000
Saintes Maries, 1974	0.95	0.45	470 000
Carro, 1974	0.68	0.20	300 000
Marseille, 1974	1.60	1.10	690 000
California:			
Los Angeles, 1972	1.20	0.23	190 000
San Francisco, 1975	0.98	0.068	69 000
DDT or DDE			
France:			
Valras Beach, 1974	0.54	0.051	90 000
Saintes Maries, 1974	0.35	0.013	40 000
Carro, 1974	0.25	0.052	210 000
Marseille, 1974	1.30	0.900	690 000
California:			
Los Angeles, 1972	5.10	1.600	310 000
San Francisco, 1975	0.11	0.005	45 000

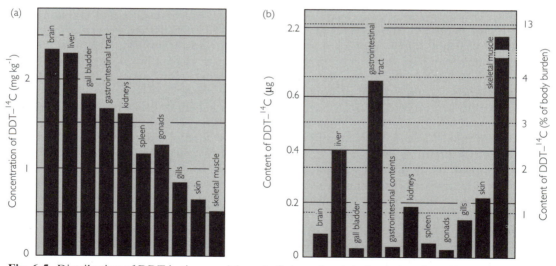

Fig. 6.5 Distribution of DDT in tissues of the sole *Solea solea*: (*a*) after being given 17 μg of DDT during three days, (*b*) after two months in a DDT-free environment.

substances in water, there is considerable uncertainty about the dose actually received by aquatic organisms in laboratory tests unless the organochlorine is administered by mouth. Because organochlorines are stored in the fatty tissues of the body, they become biologically available and exert their influence only when the fat tissues are metabolized.

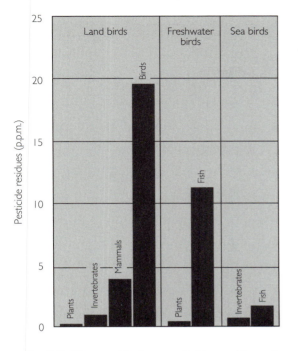

Fig. 6.6 Organochlorine pesticide residues in land, freshwater, and sea birds, in relation to their predominant diet.

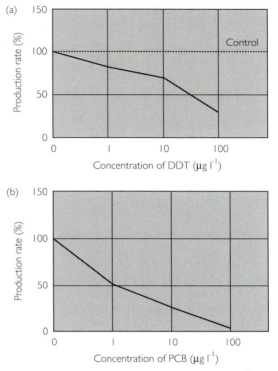

Fig. 6.7 Production (as percentage of control) of (*a*) whole phytoplankton from the western Caspian Sea exposed to DDT, and (*b*) whole phytoplankton from the Mediterranean Sea exposed to PCBs.

Animals may therefore acquire a considerable body-burden of halogenated hydrocarbons, but show no ill effects except in conditions of starvation, when fat reserves are mobilized. These facts make laboratory experimentation difficult but, if confirmation of laboratory findings is sought in natural populations, there is the additional complication that wild animals are usually contaminated with a variety of halogenated hydrocarbons and it is impossible to tell wihch one is responsible for whatever symptoms have been detected.

In laboratory cultures of whole phytoplankton from the Caspian and Mediterranean Seas, DDT and PCBs reduce primary production (Fig. 6.7) by as much as 50 per cent at a concentration of only 1 μg l^{-1} (p.p.b.) in the case of PCBs. Marine crustaceans and fish also appear to be very sensitive to organochlorines. The 96 h LC$_{50}$ of DDT ranges between 0.4 and 89 μg l^{-1}, and of dieldrin between 0.9 and 34 μg l^{-1} for

a variety of teleosts: puffer (*Sphaeroides maculatus*), killifish (*Fundulus heteroclitus* and *F. majalis*), mullet (*Mugil cephalus*), eel (*Anguilla rostrata*), silverside (*Menidia menidia*), and bluehead (*Thalassoma bifasciatum*). The values for the shrimps *Crangon* and *Palaemonetes*, and the hermit crab *Pagurus longicarpus*, are 0.6–6.0 μg l^{-1} for DDT and 7–50 μg l^{-1} for dieldrin. There is, however, a wide range of toxicity among chlorinated hydrocarbons. Figure 6.8 shows the relative toxicity of this class of substances to *Crangon* and *Astacus*. Bivalve molluscs, on the other hand, with their ability to concentrate organochlorine pesticides without coming to harm, have a 96 h LC$_{50}$ greater than 10 000 μg l^{-1}.

Since halogenated hydrocarbons are

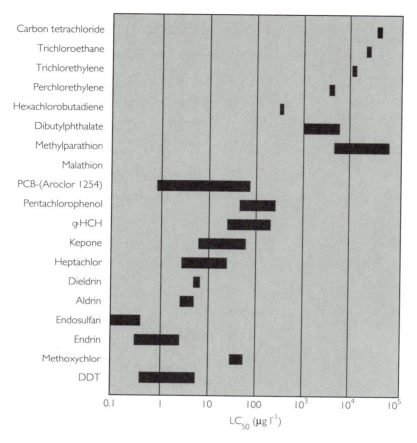

Carbon tetrachloride
Trichloroethane
Trichlorethylene
Perchlorethylene
Hexachlorobutadiene
Dibutylphthalate
Methylparathion
Malathion
PCB-(Aroclor 1254)
Pentachlorophenol
g-HCH
Kepone
Heptachlor
Dieldrin
Aldrin
Endosulfan
Endrin
Methoxychlor
DDT

0.1   1   10   100 10^3 10^4 10^5

LC_{50} (μg l^{-1})

Fig. 6.8 Toxicity of organochlorine compounds to shrimp *Crangon* and crayfish *Astacus.*

known to be subject to bioaccumulation and biomagnification, most attention has been paid to their effect on animals high in the food chain. Very large doses of PCBs and DDT amounting to several g kg^{-1} of the body weight, are required to cause the death of mammals or birds by a single administration by mouth. Such doses are clearly not found in the natural environment, where animals are exposed to much lower levels of contamination of their food, but are continuously, or chronically, exposed to them.

Two sub-lethal effects of halogenated hydrocarbons in birds and mammals appear to have some ecological significance. DDT and its residues interfere with calcium metabolism and result in the egg shells being unusually thin. In both birds and mammals, PCBs have a more direct effect on reproduction, and eggs of birds contaminated with

PCBs show reduced hatchability. PCBs included in the diet of laboratory rats and mice cause a reduction in the number of offspring produced and poorer survival of those young that are born. Mink are very severely affected in this respect when fed PCBs.

ENVIRONMENTAL IMPACT OF HALOGENATED HYDROCARBONS

Warning signals

During the 1960s there were increasing signs that the widespread and intensive use of pesticides, most obviously DDT, was having unforeseen and unwelcome consequences for the natural environment.

In 1964, fish were found to be dying in an

area around the marine outfall of a Danish factory manufacturing the pesticide parathion. Lobsters, *Homarus gammarus*, were affected over a much wider area. The factory effluent was found to be lethal to lobsters at a dilution of 1:50 000.

The Laguna Madre was a productive lagoon system on the coast of Texas, but it became heavily contaminated with pesticides from neighbouring agricultural land. The fish catch fell abruptly after 1964–5 (Table 6.5) and this was attributed to pesticide poisoning.

Table 6.5 Capture of sea trout in Laguna Madre

Year	Catch (t ha^{-1})
1964	74
1965	62
1966	30
1967	—
1968	6
1969	0.4

No sample in 1967.

The Salinas valley is one of the richest agricultural areas in California, specializing in vegetable and salad crops. The Salinas river which drains the valley does not flow in summer but is blocked by a sand bar at its mouth. During the rainy season, December–April, the river breaks through the bar and flows into Monterey Bay. Although it is wide open to the Pacific Ocean, the bay has a sluggish and irregular circulation and any pollutant remains there for some time. In exceptionally wet seasons the river may burst its banks and flood large areas of farmland. This happened in February 1969, when the river burst the sand bar and flood water carrying enormous quantities of silt flowed into the bay. During the previous 10 years, DDT had been used intensively (it is estimated that 50 t yr^{-1} had been used on crops in the valley) and pesticide residues were carried with the silt-laden flood water into the bay. This was followed by exceptionally large numbers of sea bird deaths in the following

months. All had remarkably high liver concentrations of DDT residues:

Brandt's cormorant	107–155 p.p.m.
Western grebe	199–222 p.p.m.
Fork-tailed petrel	313 p.p.m.
Ashley's petrel	412 p.p.m.
Ring-billed gull	805 p.p.m.

In addition, sea lions (*Zalophus californianus*) were found dead or dying on beaches. They had liver concentrations of DDT up to 89 p.p.m.

The dangers of a widespread ecological impact from the use of persistent pesticides were expressed by Rachel Carson in *Silent Spring*, published in 1963. This book presented the case against pesticides and if some of the predictions were overstated, it drew public attention to the problems associated with the widespread and indiscriminate use of pesticides.

Population decline of predatory birds

The most reliable evidence of the damaging effect of organochlorine pesticides on wildlife was demonstrated in 1967 by Dr D. A. Ratcliffe of the Nature Conservancy in the United Kingdom. The peregrine falcon (*Falco peregrinus*) was protected in Britain after 1945 and showed a dramatic increase in numbers until, in the mid-1950s, the population went into a sharp decline. This proved to be due to reproductive failure: birds were laying eggs with abnormally thin shells and a large proportion of them were broken during incubation. Examination of museum specimens showed that egg-shell thinning had started about 1947. Thinning of egg shells can be produced in ducks and chickens if DDT is included in their diet. High concentrations of DDT residues occurred in peregrines and the yolk of their eggs during the mid-1960s. There was no doubt that DDT was the cause of the population decline of these birds and, with the cessation of the use of DDT for agricultural purposes in Britain, peregrine numbers have increased to their former level.

Pesticides have persisted in the environment, however, and for some reason there has not been a marked decline in DDT residues in sparrow hawks (*Accipiter nisus*) in Britain. This species appears to have been more affected by the cyclodiene organochlorines, and the recovery of its population was delayed until the 'drin' pesticides went out of use in 1975.

The peregrine falcon is a top predator and therefore particularly vulnerable. Comparable effects of organochlorine pesticides were not found in herbivorous or insectivorous birds, but other predators were affected. The North American sparrow hawk (*Falco sparverius*) is related to the peregrine and suffered a similar fate. A number of fish-catching predators were also affected, including the American bald eagle (*Haliaetus leucocephalus*), osprey (*Pandion haliaetus*), and brown pelican (*Pelicanus occidentalis*), all of which suffered a serious population decline. Pelican colonies on the Californian coast had their first successful breeding in 1972 after nearly 10 years of reproductive failure, following curtailment of the use of DDT as an agricultural pesticide. The situation in these pelicans, however, was complicated by the scarcity of sardines, their staple food, during the breeding seasons of 1969–72 and it is not known how far this contributed to their reproductive failure.

Smaller, fish-eating sea birds appear not to have been so seriously affected by organochlorine pesticides. One example to the contrary is of Sandwich terns, *Sterna sandvicensis*, and eider duck, *Somateria mollissima*, on the island of Griend off the coast of Holland. It is thought that effluents from a chemical factory near Rotterdam producing dieldrin, endrin, and telodrin, were responsible for a decline in the size of the tern colony from 20 000 to 650 in 1965. There were many eider deaths between 1964 and 1967 when measures were taken to prevent the discharge of effluent from the factory. By 1974, the numbers of breeding terns had risen to 5000.

The difficulty in interpreting sea bird deaths is illustrated by the wreck of sea birds (mainly guillemots, *Uria aalge*) in the Irish Sea in autumn 1969. Over 12 000 birds came ashore in a dying condition in the space of a few weeks. Some estimates put the death toll at 50 000 or even 100 000 birds. All were emaciated and those autopsied had liver and kidney lesions similar to those caused by PCBs. While the total body-burden of PCBs was about the same in the affected birds as in other, apparently healthy, birds, very much higher concentrations were recorded in the liver. This suggests a redistribution of chlorinated hydrocarbon within the body resulting from the loss of fat in these emaciated birds. These results are based on a small sample of birds which proved to be highly variable, with PCB concentrations in the liver ranging from 8–880 μg g^{-1}. Although PCBs may have been a contributory cause of death in some birds, it cannot be concluded that poisoning by chlorinated hydrocarbons caused the wreck. Furthermore, sea bird wrecks were recorded from time to time in the last century, long before the introduction of these compounds.

Impact on seal populations

Organochlorines, or more particularly PCBs, have been held responsible for the decline of some seal and sea lion populations, though the evidence is far from complete and there are a number of unexplained discrepancies.

Seal populations in the Baltic have been declining since the beginning of this century. Much of the decline was due to over-hunting, but it has continued despite protection of the seals and there is a strong suspicion that high levels of contamination by PCBs has been responsible for a failure of reproduction. Since about 1970, only 28 per cent of mature female ringed seals (*Phusa hispida*) have become pregnant each year, instead of the usual pregnancy rate of 80 per cent. Non-pregnant females contain an average of 77 p.p.m. PCBs in the blubber lipids, compared with 56 p.p.m. in pregnant females. About 40 per cent of females have one or both uterine horns blocked by occlusions, effectively rendering them sterile. Grey seals (*Halichoerus grypus*) and common, or harbour seals (*Phoca vitulina*) found dead, showed the same pathological conditions; non-pregnant females contain 110 p.p.m. PCBs and pregnant females 73 p.p.m. PCBs in the blubber lipids.

PCBs are known to interfere with ovulation and development in mammals, though this has not been confirmed experimentally in seals. PCB concentrations up to 145 p.p.m. in Californian sea lions (*Zalophus californianus*) have been suggested as a cause of abortions, but these specimens had exceptionally high concentrations of DDT residues as well (up to 5077 p.p.m. in the blubber) and cannot be compared with Baltic seals. PCB contamination may explain why the pregnancy rate of Baltic seals without occlusions is only 50 per cent, but there is no evidence to suggest that PCBs are responsible for the uterine occlusions which render 40 per cent of the seals infertile. Furthermore, seals on the Dutch coast and on the Farne Islands in northeast England, with higher concentrations of PCBs than those in the Baltic, do not show this pathological condition. The birth rate of the Farne Island population of grey seals, despite up to 122 p.p.m. contamination with PCBs, grew steadily until the early 1970s, when attempts were made to control the population (Fig. 6.9).

In 1988, an epidemic of viral distemper severely reduced the population of common seals in the North Sea. The epidemic started in seal colonies on the Danish and Dutch coasts and then spread to British colonies. Grey seals were less seriously affected. There has been some speculation that the establishment of the virus in the seals was facilitated in populations already weakened by pollution, but there is no firm evidence of this and lethal epidemics have been observed in seal populations in other parts of the world at various times.

PERSISTENCE OF CONTAMINATION

Because of the long life of chlorinated hydrocarbons in the environment, their effect may continue for some years after an input has been brought under control.

Until 1971, the ocean outfall of the Los Angeles sewerage system carried a large quantity of DDT residues from a factory manufacturing the insecticide. Offshore sediments were heavily contaminated over a wide area (see Fig. 6.2) but, despite a dramatic reduction in DDT emissions after 1971, contamination of fish in the area persisted for years. Table 6.6 shows the modal concentration of DDT residues in the muscles of three species of fish sampled twice a year in the outfall areas, and also the percentage of the fish in the sample with more than the recommended limit for human consumption of 5 p.p.m. Dover sole (*Microstomus pacificus*), which feeds on invertebrates living in the bottom sediment, actually contained more DDT residues (median 14 p.p.m.) in 1975 than in 1971 (4.7 p.p.m.) and throughout the period, in most samples, more than 80 per cent of the fish contained more than 5 p.p.m. Black perch (*Embiotoca jacksoni*) is also a bottom feeder, but takes organisms attached to rocks. Peak median concentrations of 24–5 p.p.m. were recorded in autumn 1972 and spring 1973; thereafter the contamination level fell, to 3.2 p.p.m. in spring 1977. Even so, in the last sample, 20 per cent of the fish were above the permissible level of 5 p.p.m.

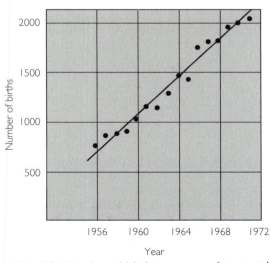

Fig. 6.9 Number of births per year of grey seals in the Farne Islands colony, 1956–72.

Table 6.6 Total DDT concentration (mg kg^{-1} wet wt muscle) of fish collected in the ocean discharge area, Los Angeles

	1970	1971		1972		1973		1974		1975		1976		1977
	Spring	Spring	Autumn	Spring	Autumn	Spring	Autumn	Spring	Autumn	Spring	Autumn	Spring	Autumn	Spring
Dover sole (*Microstomus pacificus*)														
Median	4.7	5.6	17	8.1	6.2	25	13	12	12	14				
% with more than 5 p.p.m.	50	80	86	80	67	100	85	82	82	83				
Black perch (*Embiotoca jacksoni*)														
Median	8.0	18	9.1	9.3	2.5	24	4.5	6.8	5.7	6.3	2.9	4.0	3.3	3.2
% with more than 5 p.p.m.	75	100	71	73	75	92	46	59	63	55	44	23	18	20
Kelp bass (*Paralabrax clathratus*)														
Median	3.3	2.6	1.8	4.6	12	8.3	2.6	3.2	2.0	3.8	7.7	0.26	0.84	3.4
% with more than 5 p.p.m.	33	33	0	40	100	73	33	7	33	45	100	0	0	40

Kelp bass (*Paralabrax clathratus*) feeds on fish and invertebrates in mid-water. Its contamination was generally less than that of the bottom-feeding species, though more variable. Even as late as spring 1977, 40 per cent of the samples were above the 5 p.p.m. limit.

None of these three species are fished commercially near the contaminated area although they are taken by sport anglers and no doubt eaten. Harmful effects of this contamination appeared in an unexpected quarter: in the Los Angeles Zoo. In early summer 1976, within four to five weeks, all but one of the collection of Brandt's cormorant (*Phalacrocorax penicillatus*) and guanay cormorants (*P. bougainvillii*) and all the Californian gulls (*Larus californicus*) died. The birds all showed loss of appetite and then, a few days later, tremors culminating in convulsions and death. On autopsy it was discovered that the gulls had 3100 p.p.m. DDT residues in the liver and the cormorants 750 p.p.m. The brain tissue of each contained 430 p.p.m. and 220 p.p.m., respectively. This is similar to the concentration in the brain of kestrels that died in convulsions after receiving an experimental diet containing DDT for 18 months. The cormorants and gulls had been in the zoo for six to seven years and it transpired that the fish on which they were fed, queen fish (*Seriphys palitus*), came from a fishery in the Los Angeles area and contained 3.1–4.2 p.p.m. DDT residues. Brown pelicans in the zoo, also fish eaters, were unaffected, but they were habitually fed surf smelts (*Hypomesus pretiosus*) which were caught elsewhere and contained only 0.018 p.p.m. DDT residues.

THREAT TO HUMAN HEALTH

Dioxins are commonly held to be among the most toxic chemicals known. It is true that the LD$_{50}$ for guinea pigs is 1 μg kg^{-1} body weight, an extraordinarily low figure, but the lethal dose for other rodents is much higher; for hamsters the LD$_{50}$ is 5000 μg kg^{-1} body weight. TCDD results in fetal loss or abortion in rats at doses of 1–10 μg kg^{-1} body weight, and causes developmental abnormalities in mice, such as cleft palate and liver deformities.

Despite many claims to the contrary, the evidence that dioxins are seriously damaging to humans is surprisingly inconclusive. Severe occupational or accidental exposure, such as that following the Seveso incident, when an explosion at a pesticide factory showered the surrounding area with dioxins, resulted in chloracne, minor but reversible nerve damage, and some impaired liver function in the most exposed casualties, but there were no more serious consequences. Studies in Sweden and the United States have suggested a link between dioxins and soft-tissue sarcomas, but these cancers have been very rare and other studies have not confirmed the link. Similarly, claims of adverse effects on reproduction among Vietnam veterans exposed to dioxins in herbicides have not been confirmed, and no reproductive ill effects were observed among the Seveso victims.

In 1968, rice oil contaminated with PCBs caused an outbreak of 'yusho' disease in Japan. In all, 1200 people were affected and suffered darkening of the skin, enlargement of the hair follicles, and eruptions of the skin resembling acne. A majority developed respiratory difficulties which persisted for several years. The rice oil was found to contain 2000–3000 p.p.m. of a 48 per cent chlorinated hydrocarbon and the minimum dose that produced symptoms of the disease was about 0.5 g of PCBs ingested over 120 days, or about 0.07 mg kg^{-1} body-weight per day. Similar symptoms to those of yusho disease have been observed in workers at a Japanese condenser factory. This was thought to be due to local contact with PCBs and the symptoms disappeared when the use of PCBs ceased. It is not entirely clear, though, that PCBs alone were responsible for the yusho incident because workers in Finland occupationally exposed to PCBs and having high concentrations of PCBs in their blood and body fat showed no sign of adverse effects. There have been no confirmed cases of illness

resulting from ingesting PCB residues from marine organisms used as food.

Even before the outbreak of yusho disease in Japan, due to the demonstration of harmful effects of DDT on top predators among birds there was naturally concern about human exposure to organochlorine pesticide residues in food. Although there was no evidence of harm to humans from this source, legal limits to contamination of foodstuffs in commerce were introduced in a number of countries as a precautionary measure. In the United States this was set at 5 p.p.m. DDT residues and 0.3 p.p.m. dieldrin. There was naturally consternation when it was discovered that human milk of some nursing mothers in Wisconsin (and no doubt elsewhere as well) exceeded that limit for DDT. Had it been cow's milk, it would have been condemned as unfit for human consumption.

Controls on the use of chlorinated hydrocarbon pesticides quickly followed in Europe and North America. Dieldrin and related insecticides were banned or phased out in the early 1970s and DDT has been subject to a progressive withdrawal, particularly for use in agriculture.

This reduction in the use of chlorinated hydrocarbon pesticides was primarily to protect predatory birds and seals which were certainly damaged by them. These substances had not been known to have caused any human deaths, but it was obviously a wise precaution to reduce human exposure to them. However, the replacement of organochlorines by other pesticides did not have a happy record. In 1970, the progressive withdrawal of organochlorine pesticides in the United States reached North Carolina, where the use of DDT on tobacco crops was prohibited for the first time. Parathion, an organophosphorus pesticide which is not persistent but is a potent nerve poison for humans, was widely substituted for DDT. In that season there were 40 human deaths in North Carolina from the careless use of parathion, for which it is necessary to use face masks and protective clothing. Even today, deaths are regularly reported in Nigeria from the careless use of the pesticides which have replaced the organochlorines.

Dieldrin, DDT, and HCH, much of them manufactured in the developed countries, continue to be used on a large scale in the developing world, particularly in India and China. As a result, HCH residues in human milk are higher in India than anywhere else in the world. Whether or not this is harming the population has not been recorded.

7

RADIOACTIVITY

NATURE OF RADIOACTIVITY

The nucleus of an atom contains **nucleons**, which are either positively charged **protons** or electrically neutral **neutrons**, bound together by powerful nuclear forces which overcome the electrostatic repulsion between the protons. The nucleus is surrounded by a cloud of orbiting **electrons**, each of which carries a negative charge equal to the positive charge on a proton. Normally, the negative charges on the electrons balance the positive charges on the protons, and the number of electrons, and hence the charge on the nucleus, determines the element to which the atom belongs and its chemical properties (Fig. 7.1).

An atom commonly loses or gains one or more electrons in the course of a chemical reaction, or through physical processes. It then becomes a positively or negatively charged **ion** and is chemically much more reactive than the electrostatically neutral atom. The process by which this occurs is **ionization.**

Atoms of the same chemical element have the same number of protons in the nucleus, but the number of neutrons may vary. These variants, known as **isotopes**, have the same chemical properties but differ in their nuclear mass. Thus, 99 per cent of naturally occurring carbon has a nucleus containing 6 protons and 6 neutrons and is designated carbon-12 or ^{12}C because its nucleus contains 12

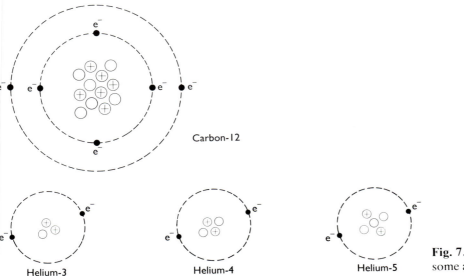

Carbon-12

Helium-3   Helium-4   Helium-5

Fig. 7.1 Structure of some atomic nuclei.

nucleons; but 1 per cent contains 7 neutrons and is designated carbon-13. Both of these are stable isotopes, but some other isotopes are unstable. Potassium with 19 protons in the nucleus normally occurs in a form with 20 neutrons as potassium-39; potassium-40, with 21 neutrons, also occurs naturally, but is unstable. The instability of the nucleus is remedied by a change in the ratio of protons to neutrons, accompanied by the emission of particles and energy. This is known as **radioactivity**, and the unstable forms are **radioisotopes** or **radionuclides**. Each radioisotope has its characteristic nuclear change and form of emissions.

α radioactivity

The unstable nucleus emits an **α particle** consisting of two protons and two neutrons. Since the nucleus loses two positive charges, the atom becomes that of an element two places lower in the periodic table. α particles are relatively slow moving and lose their energy in a short distance; they are stopped by a few centimetres of air or only 40 μm of tissue. They are, however, intensely ionizing in the matter through which they pass and can cause more damage to living tissue than particles with a longer path. Nuclei emitting α particles are therefore of biological consequence if they are taken into the body, for example by ingestion or inhalation.

β radioactivity

A neutron in an unstable nucleus spontaneously changes into a proton, or vice versa, and the resulting atom is that of an element one place higher or lower, respectively, in the periodic table. To conserve electrostatic charges, a **β particle** is emitted. This is a negatively charged **electron** if a neutron has changed into a proton, or its positively charged equivalent, a **positron**, if a proton has changed into a neutron. β particles vary widely in their energy but lose most of it within a relatively short distance and can be screened by a few millimetres of perspex or 40 mm of tissue. Like α particles, their

biological significance is greatest if a β-emitter is taken into the body.

Spontaneous fission

Nuclei of some very heavy unstable elements always have a large excess of neutrons. The nucleus breaks into two large fragments representing elements in the middle of the periodic table, accompanied by a few free neutrons. This is known as **spontaneous fission** and the fission products are themselves unstable. Neutrons are slowed down only by collisions with other nuclei and, because of the infrequency of this, neutrons penetrate matter for a considerable distance. Although they carry no electric charge and do not cause ionization, the nuclei with which they collide cause intense ionization over a short distance (like α particles) and the nucleus that finally absorbs the neutron shows strong γ radiation.

γ radiation

γ rays are similar to X-rays and, like them, are deeply penetrating and strongly ionizing. Living tissues need to be shielded from γ radiation by a considerable thickness of heavy material such as lead or concrete to absorb the radiation. In addition to being emitted by nuclei bombarded with neutrons, some of the energy released by α- and β-emitters, particularly the latter, is in the form of γ rays.

UNITS

The older units for measuring radioactivity have now been replaced by new SI units, although the latter have not yet come into universal use.

Becquerel (Bq)

Radioactivity is measured by the frequency with which radioactive disintegrations of nuclei take place in a substance. The becquerel is one nuclear disintegration per second. High levels of radioactivity may reach terabecquerels (1 TBq = 10^{12} Bq). For many purposes, the level of radioactivity is more significant than the mass of radioactive material; it is then usual to express the total amount of the substance in terms of the number of becquerels it contains.

Curie (Ci)
The former unit of radioactivity, now replaced by the becquerel, was the curie. It is defined as the amount of radioactivity displayed by 1 g of radium (^{226}Ra), that is 3.7×10^{10} Bq. It is a very large unit; the radioactivity of seawater, for example, was measured in picocuries (1 pCi = 10^{-12} Ci).

Half-life
The radioactivity of a substance declines with the passage of time. After one half-life, the radioactivity is halved. Each radionuclide has its characteristic half-life which may be a fraction of a second, hours, days, months, or millions of years. The half-life of ^{226}Ra is 1602 years. Radioactivity is inversely related to the half-life and a substance with a long half-life has low radioactivity.

Gray (Gy)
The becquerel takes no account of the nature of the disintegration, merely its frequency. For biological purposes it is more important to know the radioactivity absorbed by a tissue or an organism. This is measured by the gray, defined as the amount of radiation causing 1 kg of tissue to absorb 1 J of energy.

Rad
This unit has been superceded by the gray. (1 Gy = 100 rad.)

Sievert (Sv)
Different kinds of radiation cause different amonts of damage to living tissue for the same energy, and neutrons or α particles have about ten times the effect of β or γ particles for the same number of grays. The sievert is an arbitrary unit designed to take account of the difference. Thus, a dose of 1 sievert could be made up of 1 gray of γ particles or 0.1 gray of neutrons.

Rem
This unit has been replaced by the sievert. (1 Sv = 100 rem.)

INPUTS OF RADIOACTIVITY TO THE SEA

Background radioactivity
Radioactivity is a natural phenomenon. Seawater is naturally radioactive, largely due to the presence of potassium-40, but it also contains decay products of uranium and thorium, and receives a continuous input of tritium (^{3}H, the radioactive isotope of hydrogen) through the activity of cosmic rays (Table 7.1).

Heavy radionuclides have a low solubility in water and tend to be adsorbed on to particulate matter. They therefore accumulate in sediments; fine sediments, with their large surface area, tend to adsorb more than coarse sediments. Thus, while oceanic seawater has a radioactivity of about 12.6 Bq l^{-1}, marine sands have a radioactivity of 200–400 Bq kg^{-1} and muds 700–1000 Bq kg^{-1}. In parts of the world with very high natural levels of radioactivity, marine sands have correspondingly high radioactive levels. The best known examples are in Kerala in southwest India and in the provinces of Espirito Santo and Rio de Janiero in Brazil. At one popular bathing beach at Guarapari, near Rio de Janiero, the visitor is exposed to a dose rate of 20 μGy h^{-1}.

Table 7.1 Natural levels of radioactivity in surface seawater

Radionuclide	Concentration (Bq l^{-1})
Potassium-40	11.84
Tritium (^{3}H)	0.022–0.11
Rubidium-87	1.07
Uranium-234	0.05
Uranium-238	0.04
Carbon-14	0.007
Radium-228	$(0.0037–0.37) \times 10^{-2}$
Lead-210	$(0.037–0.25) \times 10^{-2}$
Uranium-235	0.18×10^{-2}
Radium-226	$(0.15–0.17) \times 10^{-2}$
Polonium-210	$(0.022–0.15) \times 10^{-2}$
Radon-222	0.07×10^{-2}
Thorium-228	$(0.007–0.11) \times 10^{-3}$
Thorium-230	$(0.022–0.05) \times 10^{-4}$
Thorium-232	$(0.004–0.29) \times 10^{-4}$

Weapons testing

Inputs of radioactivity to the sea from human activities began in the later stages of the Second World War with the explosion of the first nuclear weapons, and continued with nuclear weapons testing until the signing of the test ban treaty between the USA, USSR, and UK in 1963. Subsequent weapons tests by France in 1965 and China in 1968 made minor contributions. These nuclear weapons contained enriched uranium and plutonium, and when exploded under water or near the ground produced over 200 different fission products and isotopes. These were carried in fine dust into the stratosphere where they circled the globe many times before settling back to Earth again as fall-out. Many of the radioisotopes produced do not occur naturally; some have a very short half-life, but the longer-lived radioisotopes allow the extent and distribution of contamination from this source to be detected. The most important of these are strontium-90 and caesium-137, with a half-life of about 30 years, and plutonium-239 with a long half-life of 24 400 years.

Because of the global pattern of atmospheric circulation, most of the fall-out has occurred between latitude 45°N and 45°S, with higher levels in the northern hemisphere where most of the explosions took place. There has been a negligible input from this source since the mid-1960s (Fig. 7.2) and radioactivity from this source has been subject to natural decay ever since then. The average radioactivity of ocean water in the north Atlantic in the early 1970s, attributable to fall-out, is shown in Table 7.2, but its distribution is far from uniform. The highest levels occur in surface waters and, depending on water movements, are slowly carried to deeper water (Fig. 7.3).

Liquid wastes

Radioactive substances are included in cooling water and other liquid wastes from nuclear reactors and may be discharged into the sea.

Nuclear-powered ships and submarines

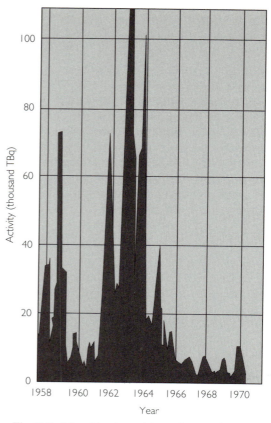

Fig. 7.2 Monthly deposition of strontium-90 in the northern hemisphere, derived from weapon testing.

Table 7.2 Average (and range of) radioactivity in north Atlantic waters in the early 1970s due to fall-out from nuclear weapon tests

Radionuclide	Radioactivity (Bq l^{-1})	
Tritium (^{3}H)	1.78	(1.15–2.74)
Caesium-137	0.008	(0.001–0.03)
Strontium-90	0.005	(0.0007–0.02)
Carbon-14	0.0007	(0.0004–0.015)
Plutonium-239		(0.0001–0.0004)

release some radioactivity, but the amounts are trivial compared with the discharges from nuclear power stations and fuel-reprocessing plants. Furthermore, such discharges as these are widely distributed in the oceans, not

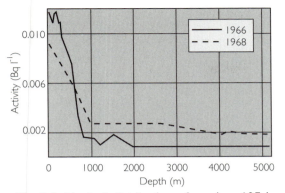

Fig. 7.3 Vertical distribution of caesium-137 in the north Atlantic.

confined to one area as in the case with discharges from land-based installations.

By 1987, 433 nuclear reactors for generating electric power were in operation or under construction in 30 countries. Several types of reactor are in use, but all depend on the energy derived from the fission of uranium, usually in the form of uranium dioxide. The core of the reactor containing the fuel rods generates great heat, and a coolant passes through the core to maintain a uniform internal temperature. The coolant may be light or heavy water, carbon dioxide, or molten sodium. The heat is transferred indirectly to water and used to generate steam to drive turbines, which generate electricity in the conventional way. This water is then condensed and recycled, but large volumes of cooling water are required for this process and are then discharged into the sea in the case of coastal or estuarine nuclear power stations.

The discharged cooling water inevitably acquires some radioactivity and the discharge also contains liquid wastes with low radioactivity from processes involved with the handling of spent fuel rods from the reactor. In Britain, the total radioactive discharge and the discharge of particular radionuclides are limited by regulations for each nuclear power station, depending on local circumstances. Thus the nuclear power station at Bradwell on the Blackwater estuary in Essex is permitted to discharge 7.4 TBq yr^{-1} of radioactivity plus 55.5 TBq yr^{-1} of tritium but, because of the presence of commercial oyster beds in the area, the discharge of zinc-65 (which the oysters accumulate) is limited to 0.185 TBq yr^{-1}. At Hinkley Point on the southern shore of the Bristol Channel, where there is no such commercial interest, levels are limited to a total discharge of 7.4 TBq yr^{-1} and 74 TBq yr^{-1} of tritium, with no stipulation about zinc-65. In practice, actual emissions are a good deal less than the permitted levels. In 1986, for example, Bradwell nuclear power station discharged 10 per cent of the permitted total radioactivity and only 1 per cent of the permitted zinc-65. Other power stations have a similar performance.

The fuel elements remain in the reactor for two to three years, by which time they have lost efficiency but still contain large amounts of uranium-235 which has not yet undergone fission, together with a variety of other radionuclides, including plutonium-239 and americium-241. Spent fuel rods may be removed for permanent safe storage on land, as is the current practice in Canada and the United States, or more commonly are taken to a reprocessing plant where the uranium is recovered for reuse and the plutonium and fission products are extracted. Major reprocessing plants in Europe are at Sellafield on the northeast coast of England and at Cap la Hague, near Cherbourg on the French Channel coast. There are other reprocessing plants at Dounreay on the north coast of Scotland, Mercoule in France, which discharges into the river Rhone, and at Karlsruhe in Germany. Reprocessing plants also exist in India, Japan, and the United States of America. Reprocessing plants discharge large quantities of waste water with a low radioactive content but, because of the quantities involved, they dwarf the output from nuclear power stations. Although the discharges from all the European reprocessing plants have been carefully monitored, more is probably known about the impact of the discharges from Sellafield than about any other source of man-made radioactivity.

Discharges from Sellafield rose to a peak in the early 1970s when the maximum permitted release of α-emitters was 222 TBq yr^{-1} and of β-emitters 11 000 TBq yr^{-1}. Since that time, discharges have been greatly reduced and by 1986 the maximum permitted discharge for α-emitters was 14 TBq yr^{-1} and for β-emitters 950 TBq yr^{-1}. In practice, actual discharges have been substantially below these limits. The most significant discharges, because of their long half-life, are caesium-137 and -134, ruthenium-106, strontium-90, and americium-241, and separate maximum permitted discharges are set for these and a number of other radionuclides. In recent years there has been some concern about the possibility of human exposure to α-emitters, primarily plutonium and americium, and discharge of these has now been reduced to a very low level (Fig. 7.4).

The behaviour of these radionuclides in the sea depends on their chemical form and physicochemical characteristics. Caesium-137 remains in solution and has a half-life of 30 years. Since the isotope does not occur naturally, it can serve as a marker for Sellafield liquid emissions, showing their course and dilution in the Irish Sea (Fig. 7.5(a)) and round the north of Scotland into the North Sea (Fig. 7.5(b)).

Other radionuclides, particularly those of ruthenium and plutonium, tend to adsorb strongly on to particulate matter and so are carried to the seabed where they accumulate in fine substrata. Some seabed sediments in the area influenced by the Sellafield discharge are contaminated with up to 3.9 Bq g^{-1} of plutonium-239. Resuspension of particles and the local hydrography results in some accumulation of radioactive particles in estuarine muds and salt marshes on the Cumbrian and Solway coasts and this has to be taken into account in calculating the radiation exposure of fishermen in the area who regularly work on the mud-flats.

Chernobyl

The accident to the nuclear reactor at

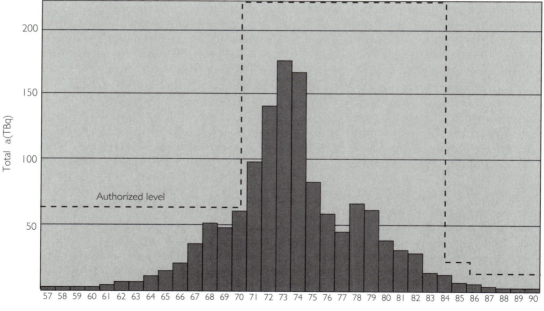

Fig. 7.4 Permitted and actual discharge of α-radioactivity to the sea from the Sellafield fuel-reprocessing plant. (*Published with the permission of the Controller of Her Majesty's Stationery Office.*)

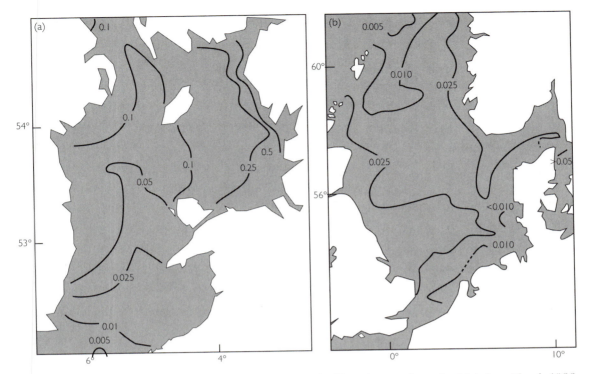

Fig. 7.5 (*a*) Concentration (Bq kg⁻¹) of caesium-137 in filtered water from the Irish Sea, March 1988. (*b*) Concentration (Bq kg⁻¹) of caesium-137 in filtered surface water from the North Sea, August–September 1988.

Chernobyl in April 1986 resulted in radio-active fall-out over considerable areas of Europe, particularly where there were heavy rainstorms as the nuclear cloud passed over. The sea area most affected was the Baltic Sea, although the North Sea, northwest coast of Scotland, and Irish Sea also received a significant input. The dominant radionuclide was caesium-137, as with routine discharges from nuclear installations. Land and fresh-water environments have more limited dispersal rates than the sea and a number of land areas remained significantly contamin-ated five years after the accident; the fall-out had a very small effect in the sea.

Seawater samples were collected all over the Baltic Sea by the German Hydrographic Institute during October and November 1986 and the mean concentration of caesium-137 was then 0.19 Bq l⁻¹ and of caesium-134 was

0.095 Bq l⁻¹. The average caesium-137 content of Baltic water before the accident, in the period 1980–5, was 0.018 Bq l⁻¹.

During the summer following the accident, fish from the western part of the Bothnian Sea contained an average of 64 Bq kg⁻¹ of caesium-137 and 31 Bq kg⁻¹ of caesium-134. In Finnish waters, fish samples collected between May and November 1986 averaged 41 Bq kg⁻¹ of caesium-137 and 19 Bq kg⁻¹ of caesium-134. Much lower concentrations were recorded in fish samples from the southern Baltic Sea.

Assuming an annual consumption of 100 kg of Baltic fish per person per year, the most critically exposed group (see p. 114) would have received a dose of less than 0.08 μSv yr⁻¹. This is only one-thirtieth the exposure of the most critically exposed consumers of freshwater fish affected by the accident.

Solid wastes

Sea dumping of radioactive solid wastes has
been practised since 1946, but has increasingly come under international control. The
1972 London Dumping Convention, now
ratified by 91 states, regulates all dumping at
sea and this agreement prohibits the dumping
of high-level radioactive wastes. High-level
waste is now defined as that containing, per
tonne of material: more than 37 000 TBq
tritium; 37 TBq β- and γ-emitters; 3.7 TBq
strontium-90 and caesium-137; or 0.037 TBq
α-emitters with half-lives over 50 years.

Intermediate and low level radioactive
waste continued to be dumped at sea until
1982. Between 1946 and 1967, the United
States used a site 2800 m deep in the western
Atlantic; a small amount was also dumped in
deep water off the Pacific coast. Eight
European countries (Belgium, France,
Germany, Italy, the Netherlands, Sweden,
Switzerland, and the United Kingdom) used
10 sites in the northeast Atlantic until 1971,
when all dumping was centred at a site 900
km from the nearest land and 550 km beyond
the edge of the continental shelf in 4400 m of
water. The amounts of radioactivity dumped
until use of this site was suspended in 1983
are shown in Table 7.3. Japan prepared a plan
to dump low level radioactive waste in deep
water in the western Pacific, but this plan has
not been proceeded with.

The waste consists of contaminated piping,
concrete and building material, glassware,

Table 7.3 Amount of intermediate and low
level radioactive waste dumped in the northeast Atlantic

Year	α emitters (TBq)	β/γ emitters (TBq)	Tritium (TBq)
1974	15.5	40 700	
1975	28.9	1130	1100
1976	32.6	1200	775
1980	70.3	3075	3630
1981	77.7	2930	2750
1982	51.8	1830	2860
Total 1948–82	660	38 000	15 000

protective clothing, and so on, and is derived
from nuclear power stations and reactors
operated by industry and nuclear research
centres, universities, research institutes, and
radiological services in hospitals. For disposal
it is packed in concrete-lined steel drums,
either embedded in resin or bitumen, or with
a pressure equalization device. This ensures
that the container reaches the seabed intact
without imploding under the great pressures.
Ultimately, the containers will corrode and
leach their contents out into the surrounding
water, but the delay results in a loss of radioactivity as the shorter-lived radionuclides
decay, and the slow release of the contents
ensures great dilution.

As on land, ocean disposal sites are
selected to ensure that the radioactive waste
does not present a risk to the human population. At such depths there is no risk of
drums of waste being accidentally trawled by
fishermen. Bottom currents on the abyssal
plane flow slowly towards the Antarctic
where they surface. Antarctic plankton might
therefore be a route by which radionuclides
could reach the human population through
the food chain. The peak annual dose from
this source, even if dumping were resumed for
another 5 years at 10 times the previous rate,
would be less than 10^{-4} mSv arising 100–500
years after the start of dumping. The
dominant contributors to this dose are
plutonium-239 and americium-241 and the
main exposure path is the consumption of
molluscs. In hypothetical pathways including
the consumption of abyssal fish, the highest
individual dose rate is estimated to be
2×10^{-4} mSv yr^{-1}. This is an exceedingly low
individual dose rate when compared with the
ICRP recommended limit of 1 mSv yr^{-1}.

Exposure of marine organisms in and
around the dump site would be at or below the
natural background level of radiation and well
below the dose rates at which harmful effects
to individuals or populations of marine organisms have been observed.

These predictions are based on theoretical
models. It is impossible to check them
because no measurable radioactivity has yet

been released at the dump sites. Even after the first drums of waste begin to disintegrate, it will be difficult to measure an increase in radioactivity against background levels because of the slow rate of discharge and dilution of soluble radionuclides and adsorption of others on to bottom sediments.

ECOLOGICAL IMPACT OF RADIOACTIVITY

It is difficult to identify the ecological impact of radioactivity in the sea. Toxicity tests on marine organisms are not very informative. Most of the field investigations have been conducted in relation to the discharge from the Sellafield reprocessing plant and, since the Sellafield discharges have historically been among the largest in the world, the experience there gives a good indication of the ecological impact of radioactivity in the sea.

Lethal doses

The measurement of acute lethal doses of radioactivity is complicated by the fact that damage inflicted on test organisms may not be revealed for some time after exposure to radiation, and sub-lethal genetic damage cannot be detected until the next or later generations. This is less of a problem for micro-organisms with a very short lifespan but, for these, measurement of the LD_{50} is hardly appropriate and, instead, the LD_{90} is used. In practice this is a measure of the dose at which almost the entire culture dies. For macroscopic animals, the LD_{50} is based on the mortality occurring within a given time after exposure to radiation. This time is arbitrarily selected, but 30 days is a convenient period and most animals surviving so long stand a good chance of living an appreciable time longer.

The LD values given in Table 7.4 are therefore intended only to give an impression of the susceptibility of different organisms to radiation damage. There is wide variation but, as a rule, the more advanced and complicated the animal, the lower the radiation dose

Table 7.4 Acute lethal radiation doses for adults of marine organisms

Organism	Dose (Gy)	Type of experiment
Blue-green algae	4000–12 000+	LD_{90}
Other algae	30–1200	LD_{90}
Protozoa	<6000	LD_{90}
Mollusca	200–1090	LD_{50}
Crustacea	15–566	LD_{50}
Fish	11–56	LD_{50}

LD_{50} values are based on total mortality observed during 30 days following exposure to radiation.

necessary to cause it fatal damage. Gametes and larval stages are more susceptible to radiation damage than adults.

Bioaccumulation and food webs

Radionuclides behave chemically in the same way as their non-radioactive naturally occurring isotopes, but the possibility of bioaccumulation and biomagnification in food chains has greater significance if the substance accumulated is radioactive.

Algae are able to acquire large concentrations of substances from the surrounding water. *Porphyra umbilicalis*, since it is the only seaweed eaten in quantity by sections of the human population in Britain, has attracted particular attention. In the neighbourhood of Sellafield, where it is exposed to a considerable variety of radionuclides from the marine outfall from the reprocessing plant, it accumulates 10 times the concentration of caesium-137 found in the water, 400 times the concentration of zirconium-95 and niobium-95, 1000 times the concentration of cerium-144, and 1500 times the concentration of ruthenium-106. Concentrations of radionuclides in *Porphyra* collected near Sellafield at the period of maximum discharges in 1974 are shown in Table 7.5. Other seaweeds in the area also accumulate radionuclides, but different elements predominate: *Enteromorpha* and *Ulva* accumulate much higher concentrations than *Porphyra* of zirconium, niobium, and cerium, and also accumulate plutonium-239. Fucoids, although

Table 7.5 Radioactivity of *Porphyra* collected near Sellafield in 1974

Radionuclide	Concentration (Bq g^{-1})
Ruthenium-106	8.29
Cerium-144	0.90
Zirconium-95/Niobium-95	1.37
Caesium-137	0.093
Plutonium-239, -240	0.051
Americium-241	0.023
Strontium-90	0.016

they are rather poor accumulators of ruthenium-106, accumulate very much higher concentrations than *Porphyra* of plutonium-239 and for this reason have proved to be useful monitors of this radionuclide. Despite the high concentrations of radioactive materials in these algae, they are unaffected by them.

Planktonic crustaceans appear to be poor assimilators of radionuclides received in their food and they lose a high proportion of their intake in faeces. The fact that they moult at frequent intervals means that radionuclides adsorbed on to them are regularly shed. Planktonic crustaceans are therefore not an important pathway for radionuclides in pelagic food webs. Most inshore fish are, however, benthic feeders and in an area with a localized input of radioactive material, as at Sellafield, sedimentation processes ensure that there is an accumulation of radionuclides on the seabed with corresponding contamination of the infauna. Crabs, *Carcinus maenas*, accumulate much greater quantities of plutonium-237 than plaice, *Pleuronectes platessa*, when fed on *Nereis diversicolor* labelled with this radioisotope (Fig. 7.6); most of the radionuclide was accumulated in the digestive gland of the crab; in the plaice it was found adhering to the gut wall but not accumulated in other organs.

Bivalves are notorious for their ability to accumulate very high concentrations of metals, though specific abilities vary widely. Scallop, *Pecten maximus*, accumulate large quantities of manganese, oysters, *Ostrea*, large amounts of zinc, and mussels, *Mytilus*, iron.

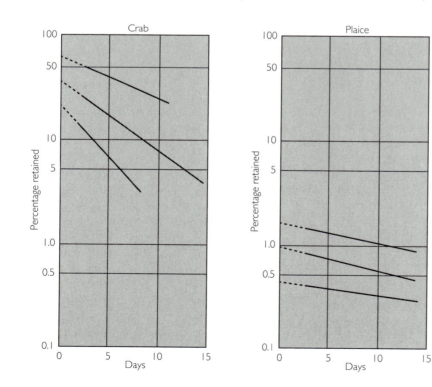

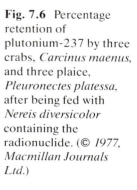

Fig. 7.6 Percentage retention of plutonium-237 by three crabs, *Carcinus maenus*, and three plaice, *Pleuronectes platessa*, after being fed with *Nereis diversicolor* containing the radionuclide. (© *1977, Macmillan Journals Ltd.*)

The very small amounts of the radioisotopes of these elements included in the discharges are readily taken up by these species.

Exposure levels

Bottom-living fish are likely to be exposed to higher levels of radioactivity than pelagic species because of the adsorption of radio-nuclides to particles that accumulate in the seabed. Experiments have been conducted to measure the exposure of plaice in the north-east Irish Sea by attaching dosimeters to their ventral surface. Individuals caught up to two and a half years after release proved to have acquired a mean dose of 3.5 μSv h^{-1}, with occasional individuals that had received a dose of 25 μSv h^{-1} (Fig. 7.7). This maximum figure is about the same as the calculated dose based on the seabed radioactivity near the Sellafield outfall. Evidently, the migrations of the fish result in the great majority of them receiving a much smaller dose. The minimum dose at which minor radiation-induced dis-turbances to physiology and metabolism can be demonstrated in the laboratory is 100 μSv h^{-1}, and it is evident that bottom-living fish, even in the most heavily contaminated areas, receive far less than such minimally damaging exposures.

Population effects

Even though marine organisms have generally a relatively high tolerance of radioactivity, it must be expected that damage to some susceptible individuals and genetic disturb-ances are an inevitable consequence of any level of radiation exposure, whether its source is natural or man-made. This would be manifested in increased mortality of adults or eggs and larvae, but in view of the enormous natural mortality this would be very hard to detect.

In one Canadian experiment, eggs and larvae of the chinook salmon, *Oncorhynchus tschawytscha*, raised in a salmon hatchery were subjected to 5–200 mGy per day, and returns to the river were measured in sub-sequent years. Large batches of eggs and young were used in the experiments and there was an increased number of abnormalities in young fish at all dose rates, but no decline in the number of adults returning to the river for spawning was ever detected.

Existing levels of radiation in the sea have so far produced no measurable environ-mental impact on marine organisms or eco-systems. This is not to say that there are no particularly sensitive organisms or eco-systems that might be adversely affected, but

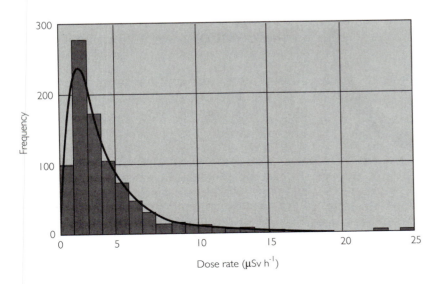

Fig. 7.7 Dose of radiation received by plaice, *Pleuronectes platessa*, in the northeast Irish Sea.

it will evidently require detailed investigations to expose them.

HAZARD TO HUMAN HEALTH

The natural environment appears to be unaffected by present levels of radioactivity, but there is naturally more concern about risks to the human population.

Exposure to radiation

The human population is exposed to radiation from a variety of sources, natural and man-made (Fig. 7.8). Natural sources are responsible for an annual dose of about 2200 μSv yr^{-1}, due to exposure to radon gas seeping from the ground, terrestrial radiation (including building material such as stone or brick), cosmic rays, and sources within the body, mainly potassium-40 ingested in food.

Artificial sources of radiation are almost wholly in medical diagnosis and therapy (chiefly X-rays). The average annual dose

from this source is about 300 μSv. Radioactive discharges, fall-out, other industrial sources, and miscellaneous make a very small contribution to the average annual exposure.

The average exposure figures conceal wide variations. Radon is particularly prevalent in granitic areas where it may accumulate in houses to a dangerous level. People living at high altitudes are more exposed to cosmic rays than those living at sea level. Medical exposure may vary from zero to near lethal doses. Exposure to 'miscellaneous' artificial sources is almost entirely due to flying.

Exposure to radioactivity from the sea is either through the consumption of seafood that has accumulated radioactive substances, or by external exposure to radioactive sediments on beaches. Some groups of people, because of their diet, work, or location, may receive higher doses than members of the general public. Table 7.6 shows the estimated maximum doses that may have been received by a selection of such critically exposed groups in northern Europe in recent years. Different groups of people are involved in each case. Massive doses of radiation result in radiation sickness and death but, at the low doses described so far, exposure must con-

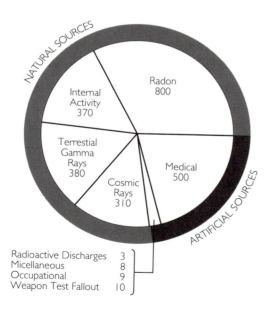

Fig. 7.8 Natural and artificial sources of radiation to which the human population is exposed in the U.K. Average annual dose in μSv.

Table 7.6 Estimated maximum doses of radiation received by critically exposed groups of people

Direct inputs into the marine environment	Annual individual effective dose equivalent (mSv)
Sellafield	0.3–3.5
Dounreay	0.03
Cap de la Hague	0.1
Other nuclear sites	0.0001–0.3*
Solid waste disposal	0.00002†
Chernobyl fall-out	0.08‡
Weapons testing fall-out	0.001–0.01
Naturally occurring radiation	< 2

*Doses towards the higher end of the range are largely due to the influence of Sellafield.
†Peak hypothetical dose in the future due to all past disposals.
‡During 1986.

tinue over a long period (chronic exposure) before adverse effects are clearly manifested.

Effects of radiation

Chronic exposure to ionizing radiation may produce **somatic** effects, most importantly leukaemia and cancer of the bone, thyroid, lungs, or breast; or **genetic** damage manifested in some abnormalities or defects in the next generation, resulting from irradiation of the gonads. While these effects undoubtedly occur, it is always difficult to establish a connection between low level chronic exposure and the incidence of an effect in any individual. Such diseases do not usually reveal themselves until long after the period of exposure; there is not a strict relationship between the intensity and duration of exposure and the incidence of disease or effects in the individual at risk; and all these diseases occur in any normal population, whether or not it is exposed to a particular source of radiation. Equally, genetic effects show a dose relationship only at the level of the population and the consequences of exposure for the individual are unpredictable. The nature of the genetic damage resulting from radiation does not differ from genetic defects that occur spontaneously.

Different radionuclides present different kinds of health hazard, depending on their chemistry. In the same way as the stable isotopes, depending on the form in which radioisotopes are presented to the organism, they differ in the ease with which they are assimilated, in the organs in which they accumulate, and the length of time they remain in them. The existence of **critical organs** in which particular substances accumulate is particularly important for α- and β-emitters, which have an intensely ionizing effect but are damaging only over a short distance. The critical organ for the radioisotope iodine-131 is the thyroid because the thyroid accumulates 600 times the iodine concentration of other tissues in the body. The critical organ for strontium-90, which behaves chemically like calcium, is bone. For manganese-54 the critical organ is the liver.

A great many factors must therefore be taken into account in assessing the possible impact of radioactive effluents on a section of the population:

(1) the nature of the radionuclide;
(2) the nature of the emitted radiation;
(3) its half-life;
(4) the decay scheme;
(5) the chemical form in which the radionuclide is encountered;
(6) the fraction that is assimilated;
(7) the organ in which it may accumulate and the concentration that may be reached.

The principal source of guidance about radiation exposure is the International Commission on Radiological Protection (ICRP). This body grew out of an international commission set up in 1928 to advise on tolerable levels of X-ray and radiation exposure. There is no evidence that there is a threshold level of radiation exposure below which there are no harmful effects, and when setting tolerable levels of exposure it must be assumed that any increase in radiation levels above the inescapable natural background level, carries some risk of damage. No human activity is without some degree of risk, however, and a degree of risk is acceptable if the benefits of an activity are great enough. The use of X-rays in medical diagnosis, for example, entails some risk of radiation damage, but given the safeguards that are in use, the risk is an acceptable one.

Legal limits of exposure

Limits to radiation exposure recommended by ICRP are designed to safeguard those regularly exposed in the course of their work. The overall limit previously prescribed was 50 mSv yr^{-1}, and this figure was adopted in an EC directive in 1980 and has become the legal limit in many countries. Particular tissues should not receive more than a fraction of this dose and there are detailed recommendations about the annual exposure to 240 different radionuclides that may be ingested or inhaled.

Occupational exposure to 50 mSv yr^{-1} would, on statistical grounds, result in 340 additional deaths per million workers per year. For comparison, in the United Kingdom in 1986–7 there were 300 accidental deaths per million in metal mining, 106 per million in coal mining, 95 per million in metal manufacture, and 92 per million in the construction industry.

In practice, occupational exposure to radiation is well below the legal limit, but for some time it has been recognized that the limit is probably too high and lower limits have been introduced in several countries. In 1987, the British Health and Safety Commission recommended that the circumstances in which any worker was exposed to 15 mSv yr^{-1} should be examined. It has recently recommended that any worker exposed to 75 mSv of radiation in any 5-year period since the beginning of 1988 should also be subject to review. Although these recommendations do not have the force of law, they have generally had a similar effect.

In 1991, the ICRP revised its recommended limits to 20 mSv yr^{-1} and 100 mSv over a 5-year period and these limits are likely to be adopted in place of the previous recommendations.

Members of the general public are much less regularly exposed to radioactivity from man-made sources than those employed in nuclear installations. The ICRP has recommended that for routine exposures, excluding medical and natural sources, the dose should not exceed 1 mSv yr^{-1} for lifetime exposure, or 5 mSv yr^{-1} for short periods. In the event of accidents, such as that at Chernobyl, control of foodstuffs should be considered if exposure is 5 mSv yr^{-1} in the first year, and should certainly have been attempted if the dose is 50 mSv yr^{-1}. A whole-body dose of 5 mSv yr^{-1} is one-tenth that previously recommended as the limit for occupational exposure. It entails a risk of one to ten radiation-induced deaths from cancer per million of the exposed population. Allowing

for the fact that there is wide variation in the levels of exposure among the population, such an upper limit means that the average dose received by members of the public is about 0.5 mSv yr^{-1}. This is about the same dose as might be expected from living in a brick house rather than a wooden one.

Critical path analysis

A more stringent safeguard is provided in setting limits to radioactive discharges to the sea by directing attention to the most likely route by which radionuclides may reach the human population, and the dose likely to be received by that section of the population most at risk. This **critical path** approach is exemplified by the way in which it has been applied to discharges from the fuel reprocessing plant at Sellafield.

A large range of radionuclides is included in the liquid effluent, which is discharged from the reprocessing plant by a 2.5 km long pipeline into the Irish Sea. The Irish Sea supports a variety of commercial fisheries and the most critical pathway to man through the consumption of contaminated seafood from the area proved, surprisingly, to be by the seaweed *Porphyra*. *Porphyra* was collected along the coast near Sellafield and shipped to South Wales where, along with *Porphyra* harvested in other parts of Britain, it was processed into laverbread, a delicacy whose consumption is almost entirely limited to South Wales. *Porphyra* accumulates a variety of radionuclides from seawater (Table 7.5), the most important being ruthenium-106. Surveys carried out in South Wales showed that 26 000 people consumed up to 75 g day^{-1} of laverbread regularly, but a small group of 170 adults was identified with individual consumption rates of about 160 g day^{-1}, with a maximum of 388 g day^{-1}.

The critical tissue for ruthenium-106 is the lower large intestine, and for strontium-90, plutonium-239, and americium-241, it is bone. Because of the variety of radionuclides in the *Porphyra*, calculations were made for individual radionuclides, on the most pessimistic assumptions, ensuring that the total exposure did not exceed recommended limits.

The recommended annual dose limit recommended by the ICRP for the lower large intestine was 15 mSv yr^{-1}; estimated doses received by the critical sub-group of laverbread eaters in 1962–7 was 4–7 mSv yr^{-1}. The recommended genetically acceptable dose was calculated to be 1.3 mSv yr^{-1}; the dose resulting from laverbread consumption was 1 μSv yr^{-1}. These calculations were based on the very conservative assumption that all *Porphyra* used in the manufacture of laverbread came from the neighbourhood of Sellafield, and ignored the water added during its preparation, which dilutes the radioactive component.

Use of the critical pathway approach requires constant vigilance to take account of changing circumstances and increasing knowledge of the behaviour of radionuclides. *Porphyra* ceased to be a significant pathway in the mid-1970s when the last two commercial collectors of the seaweed retired and Sellafield *Porphyra* was no longer used in the preparation of laverbread. Locally caught fish then assumed the critical position: the maximum exposure of an individual in the critical population from fish consumption was 34 per cent of the dose limit recommended by ICRP. Caesium-137 contributes substantially to radiation exposure from eating fish caught around Sellafield, and caesium discharges rose substantially in 1974–5 as a result of changed practices at the reprocessing plant. In 1976, the critical group among members of the public received 25 per cent of the ICRP recommended dose limit of caesium-137 from this source. In subsequent years, this figure was reduced to about 10 per cent of the

ICRP limit. Further reassessment of exposures was necessary in 1982 when a new survey by the Ministry of Agriculture, Fisheries, and Food revealed that the estimated consumption of molluscs by the critical group of seafood consumers had increased threefold. Their intake of plutonium and americium had increased by a similar factor. It had also been discovered that the uptake of plutonium from food by the human gut is five times higher than was formerly thought. Thus the calculated dose of plutonium received by members of the critical group must be increased by a factor of 15, bringing their total exposure from 24 per cent to 39 per cent of the ICRP annual dose limit, of which actinides (plutonium and americium) contribute 26 per cent rather than the 4 per cent previously estimated. Steps were then taken to reduce the discharges of α-radioactivity accordingly.

Critical pathway analysis is used to determine discharge standards in a number of countries and this accounts for the great variation in the limits set for each installation, which must take account of local circumstances and the critical pathways of exposure of the human population to radioactivity. The nuclear power station at Lake Trawsfynydd in Wales discharged 5.18 TBq in 1975 and individuals in the critical group consuming fish from the lake are estimated to have received 8 per cent of the ICRP limit. The reprocessing plant at Sellafield, on the other hand, discharged 10 000 TBq in that year, when maximum exposure of a member of the critical group of seafood consumers was estimated to be 34 per cent of the ICRP limit, only four times as much despite the enormously greater discharge.

8

SOLID WASTES AND HEAT

DREDGING SPOIL AND INDUSTRIAL SOLID WASTES

Dredging spoil

Ports, harbours, rivers, and approach channels often need regular dredging to keep them open for shipping. The dredged material may be used for land reclamation, but is more often dumped at sea. Over 91 m t of dredging spoil was dumped at licensed offshore sites in northwest European waters (excluding the Baltic) in 1986. The United States dumps three times as much at sea.

The composition of the spoil is varied: for routine dredging it is usually fine silt but, if new channels are constructed or old ones deepened, it may consist of boulder clay, chalk, and so on. Unlike other solid wastes dumped at sea, the spoil cannot usually be distinguished from the surrounding substratum at the dumping ground and, depending on the bottom currents, it may not stay where it was dumped, but gradually be transported to other areas, including the place it originally came from.

Dredging spoil is often anoxic and, particularly if it comes from harbours or indus-trialized estuaries, is usually contaminated with metals, pesticides, and persistent oils, the legacy of a century or more of discharges into the rivers. These contaminants are adsorbed on to silt particles and are generally not available to organisms. Dredgings made in shipping channels at sea off the river mouth may be heavily contaminated, but material from new channels or from offshore sand-banks is uncontaminated.

These contaminants are transferred to the dumping grounds. Table 8.1 shows the quantities of heavy metals included in dredging spoils dumped by Belgium, the Netherlands, and Britain in 1986, mostly in the North Sea. These totals conceal wide variations: dredgings from harbours and estuaries with intense shipping activity and heavy industry are much more contaminated than that from other areas. Dredging spoil from docks on the Manchester Ship Canal (see p. 18), for example, contains 20.7 p.p.m. mercury and 5080 p.p.m. lead. Spoil from the Tees estuary (also with long-standing indus-trial inputs) contains 7 p.p.m. mercury and 3000 p.p.m. zinc, but only 320–460 p.p.m. lead. Spoil from Swansea docks contains a

Table 8.1 Quantities of metals (t yr^{-1}) in dredging spoils dumped in the North Sea in 1984–5

Country	Total (× 10^3 t)	Cadmium	Mercury	Copper	Lead	Zinc	Chromium	Nickel
Belgium	47 690	—	10.5	616.0	1542.0	8976	2309.0	483
Holland	28 340	22.9	8.9	341	763.7	—	749.4	219.1
UK	13 810	5.3	6.9	424.7	647.2	1565	667.6	249.1
Total*	96 115	30.2	27.1	1429.5	3092.9	13 586	3842.4	1005.9

*Including smaller quantities from France, Germany, and Denmark.

high concentration of cadmium (18 p.p.m.), perhaps from natural sources. At the other extreme, the chromium concentration in the spoil from Newhaven harbour (a ferry terminal in a non-industrialized area) is 1 p.p.m., compared with 1500 p.p.m. in one sample from the Tees estuary.

At dump sites where dredging spoil remains in piles on the seabed, it generally does not become oxygenated and the contaminants remain biologically unavailable and so do no harm. However, if the dredging spoil does become oxygenated, metals may change their species, cease to be adsorbed on to silt particles, and then enter food chains. Because of this risk, some dredgings are regarded as too contaminated to be safely dumped at sea. Dredgings from the inner harbour of Rotterdam are in this category and are no longer dumped in the North Sea, but are deposited on an artificial island. This island has a capacity for about 20 years' dredgings, by which time it is hoped that the river Rhine, the source of most of the contaminants, will have been sufficiently cleaned for sea-dumping to be resumed. Dredgings from the Elbe at Hamburg are also too contaminated to be dumped at sea and are now placed on an artificial island.

The immediate impact of dumping dredging spoil is a smothering of the benthic fauna in the dumping grounds, though some animals are able to burrow to the surface of the dumped material without much difficulty. One study was carried out in the fjord at Uddevalla on the west coast of Sweden (Fig. 8.1) where dredging was going on in an area where surveys of the benthic fauna had been made in 1971–2. Immediately after the dredging of a new channel, there was a loss of diversity at stations near the site of operations due to a failure of recruitment of several bivalves, probably because of the increased sediment in the water, but the position was restored a year later. Benthic animals in the area accumulated high concentrations of metals when dredging was in progress, but returned to former levels within 18 months.

China clay waste

China clay deposits near St Austell in south Cornwall have been used since the middle of the eighteenth century as a source of fine clay for the manufacture of porcelain. The waste which the extraction of the clay produces consists of fine sand, kaolin, and flakes of mica, and has been discharged into the rivers Fal, Luxulyan, and St Austell (Fig. 8.2), though the Fal ceased to be used for this purpose many years ago. About 1.5 m t yr^{-1} of mica waste is produced, about half of it being discharged into the St Austell river (known locally as the White river for this reason), 25 per cent into the Luxulyan river, and the rest disposed of on land.

Most of the river-borne waste is deposited in St Austell Bay and Mevagissy Bay (Fig. 8.3(a)). The substratum in the area is naturally one of coarse sand, shell, pebble, and rock but about three-fifths of the seabed is now covered, sometimes to a considerable depth, with micaceous clay waste and this has caused faunistic changes. Some bivalves, such as *Venus fasciata* (Fig. 8.3(b)) and *Dosinia exoleta*, are intolerant of the clay waste but there is a characteristic fauna of the china clay waste, including the polychaete *Melinna palmata*, the bivalves *Tellina fabula* and *Cultellus pellucidus*, the heart urchin *Echinocardium cordatum*, the holothurian *Labidoplax digitata*, and the ophiuroid *Amphiura filiformis* (Fig. 8.3(c)). Plaice, *Pleuronectes platessa*, dab, *Limanda limanda*, and sole, *Solea solea*, feed on this specialized fauna.

Fly ash

Ash from coal-fired power stations consists of large, fused lumps of clinker and a considerable amount of very fine powder. This includes 'pulverized fuel ash' from the boilers but is mainly 'fly ash' extracted from the exhaust fumes in the smoke stacks by air pollution control equipment. This fine material is composed mainly of silicon dioxide with oxides of aluminium and iron, and a variety of metals present in trace concentrations. There is some use for fly ash

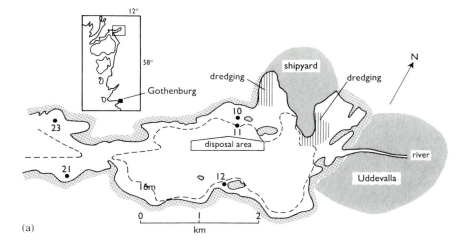

(a)

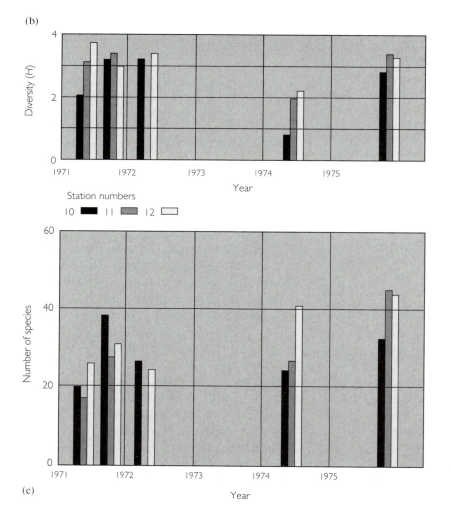

(b)

(c)

Fig. 8.1 (*a*) Dredged and disposal areas in the Byfjord, western Sweden, (*b*) diversity (Shannon–Wiener index), and (*c*) number of species at stations 10–12 before and after dredging; station 12 represents the control site. (*Pergamon Press.*)

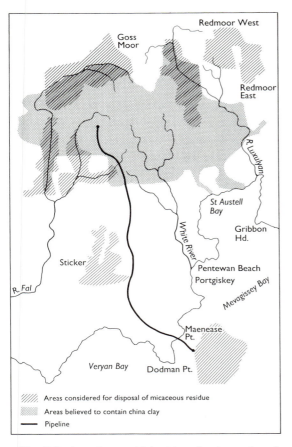

Fig. 8.2 Area from which china clay is produced near St Austell, Cornwall, and the discharge routes for the china clay waste. (*Pergamon Press.*)

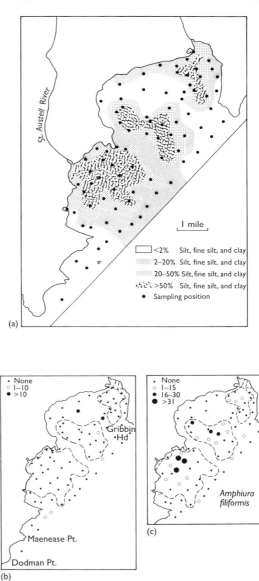

Fig. 8.3 St Austell and Mevagissey Bays, Cornwall: distribution of (*a*) clay waste fractions in bottom sediments, (*b*) the bivalve *Venus fasciata*, intolerant of clay, and (*c*) the brittle star *Amphiura filiformis*, clay-tolerant. (*Cambridge University Press.*)

in the construction industry but the supply greatly exceeds demand; the surplus is dumped generally on land, but power stations in Newcastle and Blyth in northeast England dump 600 000–700 000 t yr^{-1} at sea. The particle size of fly ash is 2–200 μm and similar to that of fine silt but, because of its homogeneity, it offers low niche diversity. At the Blyth dumping ground some species such as the gastropod *Turritella communis* and the foraminiferan *Astrorhiza*, common at the time dumping started in 1959, are now rare or eliminated. Unlike the situation at St Austell Bay with an input of clay waste, the main dumping ground receiving fly ash shows extreme impoverishment of the fauna, pre-

sumably because of the daily disturbance of dumping by barge. As in Cornwall, the fine sediment has sifted into rock crevices to the detriment of the lobster fishery, but there appears not to have been a compensating development of a rich benthic soft-bottom fauna.

Fly ash dumping at sea is now being phased out.

Colliery waste

Waste from collieries near the northeast coast of England has been dumped on beaches or barged out to sea for dumping since the start of this century. The waste is predominantly shale with some sandstone, but includes small fragments of coal and has a high metal content as a result. The waste is mainly in gravel-sized fragments or larger, 2–30 mm. Material dumped on beaches is not subsequently dispersed, but that dumped offshore is subject to considerable movement. There is great depletion of the fauna in the areas of dumping. This dumping will cease in 1993.

PLASTICS

Plastic particles

Small pellets of polyethylene, polypropylene, and sometimes polystyrene, 3–4 mm in diameter, are widespread in the oceans. Their presence has been reported on beaches and in surface waters near centres of industry or major shipping routes since the early 1960s, but they have also been found in the south Atlantic and south Pacific, several thousand kilometres from any individual source, and it must be assumed that their distribution is now worldwide. They are probably derived from plastic 'feedstock' for the plastics industry and reach the sea through accidental spillages at ports, or at factories close to rivers. They are virtually indestructible and buoyant, and are presumably steadily accumulating in the sea and on coasts.

These pellets are too small to cause an aesthetic nuisance on beaches, unlike other plastic debris. It is not certain if they are harmful to marine organisms but they are certainly ingested by surface-feeding marine birds. Prions, *Pachyptila*, are small petrels from the southern oceans, which have a sieving arrangement on the lower bill and feed on plankton. They commonly have numerous plastic pellets packed in the gizzard. Larger petrels and shearwaters rarely contain plastic pellets, but carnivorous birds such as great skuas, *Catharacta skua*, acquire them from their food and regurgitate them along with other indigestible material in their pellets. Plastic particles have also been found in the pellets of gulls and terns.

Although one can speculate that an accumulation of pellets in the gut may be harmful, the only evidence of this is a negative correlation between body fat and the number of pellets in migrating red phalaropes, *Phalaropus fulicarius*, on the Californian coast. If ingested pellets are responsible for impairing feeding, this would obviously be detrimental to birds facing a long migration.

Plastic litter

Larger plastic debris, mostly primary and secondary packaging, occurs in great abundance in the sea. Much of it is derived from shipping and it is estimated that 1.1–2.6 kg per person per day of plastic waste is generated and nearly all thrown overboard. Plastic wrappings of cargo add another 290 t per ship per year and the total amount of plastics discarded by ships may be 6.5 m t yr^{-1}, most of it within 400 km of land.

Plastics are not biodegradable and, although some may be subject to photodegradation when exposed to ultraviolet light, most are extremely durable and have a long life in the sea. Judging from plastic litter on beaches, the average age of containers is 2.9 years; older material is generally in fragments and stranded high on the shore. Over half the stranded plastic is in the form of containers for lavatory and household cleansers, which are made of polyethylene, and a substantial minority are polyvinyl chloride mineral water containers, but there is a great variety of plastic containers for other substances made from a variety of plastics.

Plastics are inert to marine organisms, and floating articles that have been at sea for some time may acquire a fauna of barnacles and other encrusting organisms. Large fish have been discovered with plastic cups lodged in their stomachs, and the blanketing of the seabed, to the exclusion of the burrowing fauna, by discarded polythene sheeting is an increasing problem on fishing grounds. Generally, however, plastic litter is an aesthetic nuisance on amenity beaches but not a hazard in the marine environment.

Drift nets

Diving sea birds, seals, and dolphins become entangled in drift nets when pursuing fish, and are drowned. Drift nets that break free continue to catch fish automatically and continuously until they are washed ashore or rot, and they are a danger to birds and sea mammals as well as to fish. This is not a new problem, but it has been greatly increased by the introduction of nylon monofilament gill nets.

Such nets are virtually indestructible and are very efficient at catching fish, perhaps because the fish cannot detect them. It is claimed that for the same fishing effort, the catch of bass, *Dicentrachus labrax*, on the south coast of Cornwall has increased from 100 kg to 3000 kg per fishing vessel per day through the use of monofilament gill nets. Unregulated use of these nets may have seriously reduced the stocks of white sea bass on the Californian coast, and the Canadian, New Zealand, and South African fishery authorities have all found it necessary to ban or closely regulate their use.

Unfortunately, the nets are equally efficient at catching sea birds. The Danish salmon drift net fishery in the north Atlantic is estimated to have killed $500\,000 \pm 250\,000$ sea birds, mostly Brunnich's guillemot, *Uria lomvia*, each year during the period 1965–75 when the fishery was exploited. The Japanese salmon fishery in the north Pacific and the Bering Sea is estimated to have taken an annual toll of between 214 500 and 715 000 birds during 1952–75, and 350 000–450 000 in 1975–8. Such losses are of an order of magnitude greater than the estimated losses

of sea birds in the north Atlantic from oil pollution.

Monofilament gill nets trap a considerable variety of fish, birds, and sea mammals. Northern fur seals, *Callorhinus ursinus*, dall porpoises, *Phocoenoides dalli*, sea otters, *Enhydra lutris*, and on the southeast African coast, Cape fur seals, *Arctocephalus pusillus*, have been recorded as being killed by becoming entangled in discarded or lost nets. There are reported to be 130 000 small cetaceans caught in nets each year, but the actual number may be much higher. Birds trapped by them include the laysan albatross, *Diomedea immortabilis*, fulmars, *Fulmarus glacialis*, shearwaters, *Puffinus griseus* and *P. tenuirostris*, and tufted puffins, *Lunda cirrhata*. Since the monofilament gill nets used in the Japanese northern Pacific salmon fishery are 6 m deep and may be 60 km long, their scope to inflict damage after they have broken free is very great indeed.

MUNITIONS

In the past, many countries have disposed of defective, obsolete, or surplus munitions by dumping them at sea. Accidental losses from wartime, and smaller losses of explosives and target ammunitions seaward of firing practice ranges, also remain on the seabed. The wreck of the ammunition ship *Richard Montgomery*, containing 3500 t of aerial fragmentation bombs and 1100 t of high explosives, which lies in the Thames estuary only 4.8 km from the Isle of Grain oil terminals, is but one example of a continuing major hazard.

Dumping grounds and known dangers are indicated on charts but, except for large bombs and shells, these munitions, often in a dangerous condition, are shifted from their original dumping site or lose position by water movements. A considerable number are recovered in the course of fishing or dredging operations or subsequently appear on the shore, and there are periodic reports of human injuries resulting particularly from chemical warfare agents. Munitions in the sea have no material impact on living resources.

Figure 8.4 shows the main sites of munitions inputs to British coastal waters; coastal waters of other countries harbour similar hazards but the information about them has not been analysed. A wide range of such material is recovered (Table 8.2), and a substantial proportion is in a sufficiently dangerous condition that it has to be destroyed *in situ* instead of being removed for safe disposal elsewhere.

Marine disposal of unwanted munitions is now regulated and deliberate dumping in

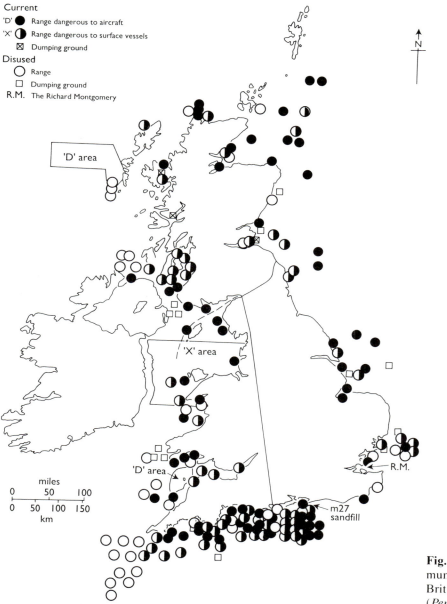

Fig. 8.4 Main sites of munitions inputs to British coastal waters. (*Pergamon Press.*)

Table 8.2 Types of munitions recovered and the proportion that had to be destroyed *in situ* in the Portsmouth and Medway Command Area (PMCA), 1969–77

Class of munitions	Number	% destroyed
Bombs	422	81.5
Land service mortars and grenades	141	37.5
Mines and torpedoes	51	26.0
Anti-submarine weapons	11	23.4
Shells	1303	21.1
Rockets and propulsion units	23	13.6
Pyrotechnics	46	6.9
Total	1997	24.4*

* Percentage of all finds destroyed.

coastal waters is not authorized except in emergency. The accidental or deliberate inputs over many earlier years, however, remain a hazard to users of the sea.

HEAT

Cooling water and often other industrial effluents are discharged at a higher temperature than that of the receiving waters. By far the greatest amount of heat discharged to the sea is in cooling water from coastal power stations. About 20 m m^3 of cooling water, 12 °C above the ambient sea temperature, are discharged for 1000 MWe of electricity generated by oil- or coal-fired power stations. Nuclear power stations are a little less efficient and the cooling water from them is about 15 °C above ambient.

In the tropics there is little fluctuation in the power demand or the sea temperature, but in temperate regions, the peak load, and so the greatest discharge of cooling water, is in winter when the sea temperature is low and falling. In subtropical areas, particularly those of North America, the peak load is in summer because of the extensive use of air conditioning. The maximum discharge of hot water is thus at a time when sea temperatures are near their maximum which may be as high as 30–35°C. Not only is the added heat dissipated more slowly than in cold seas, but the sea temperature is already near the thermal point for many organisms and, with these additions of heat, may easily exceed it.

Cooling of the heated discharge is almost entirely by mixing with the receiving water. The area affected is limited to the plume of hot water and its immediate surroundings, although the direction taken by the plume may change with changing tidal currents and so the total area under its influence is greater than appears at first sight. Even so, the outfall from the Diablo Canyon nuclear power station on the Californian coast affects only 0.7 ha; the 250 MWe Bradwell nuclear power station in Essex, discharging 2 m m^3 day^{-1} into the enclosed Blackwater estuary, raises the temperature of the surface water by only 0.2–1.7 °C. A succession of power stations using and reusing the same water along the tidal Thames have a total capacity of 8000 MWe, but do not cause a build-up of water temperature. In contrast, in subtropical waters, much larger areas may be affected: at Turkey Point, on the Atlantic coast of Florida, nearly 40 ha are affected by the cooling water discharge from a power station complex (Fig. 8.5).

It is difficult to separate the effects of hot water from other features of the discharge. Cooling water is usually treated, at least intermittently, with chlorine to discourage the

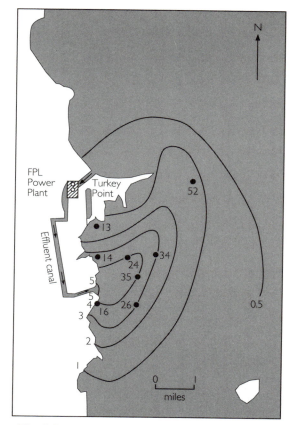

Fig. 8.5 Increase in surface sea temperature resulting from the discharge of cooling water from power stations at Turkey Point, Florida. (*Pergamon Press.*)

30–35 °C in summer and a further increase of 5 °C or more is damaging to *Thalassia*. About 9.3 ha of *Thalassia* has been destroyed by this outfall and a further 30 ha shows reduced growth.

Here and elsewhere it has been found that tropical marine animals are generally unable to withstand a temperature increase of more than 2–3 °C, and most sponges, molluscs, and crustaceans are eliminated at temperatures above 37 °C, though intertidal species appear to be more tolerant of high temperatures than benthic species. It is not surprising that there is a reduction in benthic diversity in the area of cooling water discharges from power stations on the Florida coast and in Guanilla Bay, Puerto Rico; the temperature of the discharge reaches 40–45°C in summer.

In temperate regions hot water discharges, far from being damaging, enhance the growth of bivalves and fish and advantage of this is taken for mariculture of a number of organisms. In enclosed areas receiving heated effluent, several species breed continuously and this has been observed in the ascidians *Ciona* and *Ascidiella* and the amphipod *Corophium*; others have an unusually long breeding season as in *Balanus amphitrite* and *Elminius modestus*, both exotic species to Europe. Zooplankton blooms earlier in heated water: near the Hunterston nuclear power station in the Firth of Clyde, the copepod *Asellopsis intermedia* blooms two months earlier than elsewhere in the region. Unfortunately, the phytoplankton on which the copepod depends, blooms in response to increasing day length, not sea temperature, so that the copepod does not benefit from its earlier appearance.

Strong-swimming fish are able to avoid the intake of seawater cooling systems, but smaller fish and plankton are entrained in the flow of water and organisms small enough to pass through the screens are swept through the cooling system. They may be exposed to sudden injections of chlorine, sudden and large temperature changes, and to mechanical buffeting (Fig. 8.6). There is conflicting evidence about the impact of entrainment in

settlement of organisms in the heat-exchange system. The scour of the seabed caused by the water flow in the plume is likely to change the nature of a soft substratum and so influence the fauna. Metals are leached from the cooling system and the disappearance of abalones (*Haliotis*) from the neighbourhood of the Pablo Canyon power station is attributed to copper in the cooling water, not to the elevated temperature.

The Turkey Point power station complex discharges cooling water in a situation where quiet, shallow soft-bottom areas are dominated by turtle grass, *Thalassia*. This marine grass forms the basis of a specialized community supporting a characteristic fauna. The temperature of the sea around the discharge is

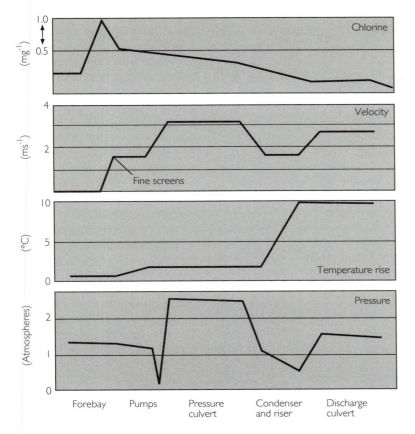

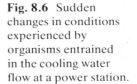

Fig. 8.6 Sudden changes in conditions experienced by organisms entrained in the cooling water flow at a power station.

the cooling water circuit. The most serious insult to entrained organisms appears to be the mechanical stresses imposed on them; small organisms, such as a polychaete larvae, appear to be little affected, but very large numbers of fish eggs and larvae may be killed.

9

THE STATE OF SOME SEAS

In this chapter, a brief view is taken of five seas: the North, Mediterranean, Baltic, Caribbean, and Caspian Seas. All receive a considerable volume of effluents but because of difference in climate, physiography, and hydrographic regimes, the problems in each differ.

THE NORTH SEA

The total area of the North Sea is 575 000 km². The southern end is constricted at the Straits of Dover, but there is no geographical northern boundary, although for practical purposes this can be regarded as a line from Shetland to the Norwegian coast near Bergen. The south and southeastern parts are shallow, with depths mostly less than 50 m. The northern part is 120–45 m deep, with deeper water off the Norwegian coast. It is not a homogeneous body of water, and a number of areas can be distinguished between which water exchange is relatively slow (Fig. 9.1). Flushing time in these areas varies between 46 and 333 days, though surface water to a depth of 10 m is exchanged more rapidly. Overall, there is an excess of precipitation over evaporation, but in winter the lee-effect of the British Isles produces a net loss of water by evaporation in the western and southwestern parts of the North Sea. During summer, all parts of the North Sea receive an excess of water by precipitation and with surface water becoming less saline, stratification of warm, less dense water over the deeper water is encouraged. A thermocline some-

times develops in the eastern North Sea and German Bight, particularly in calm weather, and this results in reduced oxygen concentrations in the bottom water (see Fig. 2.15, p. 25).

The North Sea is used as intensively, for as great a variety of purposes, as any sea area in the world. The Straits of Dover and the southern North Sea are among the most heavily trafficked sea lanes in the world; gravel

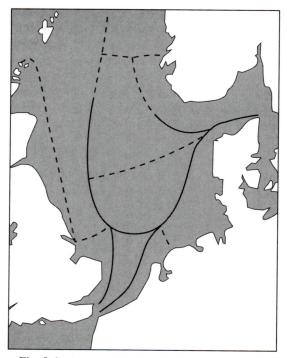

Fig. 9.1 Areas of the North Sea between which there is little water exchange.

and sand for the construction industry is dredged in a number of areas; gas and oil are extracted in the central and northern North Sea. A largely urban and industrialized population of 31 million lives around the North Sea and its estuaries, and to these are added an influx of summer visitors, particularly to the sandy beaches. The North Sea receives the wastes from this population, together with those carried in a number of major rivers: the Rhine, Elbe, Weser, Scheldt, Ems, Thames, Trent, Tees, and Tyne. It also provides the site of highly productive fisheries which are intensively exploited.

Inputs

Much of the sewage from the population surrounding the North Sea is discharged untreated or given only primary treatment, although an increasing number of towns are introducing secondary treatment. Sewage sludge is discharged from the Netherlands coast and dumped at sea by the United Kingdom, though this practice will end by 1995. These sources, together with the very large input from rivers, contribute over 1 m t yr^{-1} of BOD to the North Sea (Fig. 9.2(a)). Rivers are also a major source of the 1.5 m t yr^{-1} of nitrogen and 1 m t yr^{-1} of phosphorus input, much of it derived from run-off from agricultural land (Fig. 9.2(b, c)).

Estimates of inputs of heavy metals to the North Sea vary widely (see Table 5.1, p. 65). It is clear, though, that the atmosphere, rivers, and dredging spoils are the major contributors and that metal inputs from sewage sludge and industrial wastes are small in comparison. It is uncertain how much of the contaminated dredging spoil (see Table 8.1, p. 116) is a permanent addition to the sea and how much is recycled back into the estuaries and shipping channels from which it was dredged. The very heavily contaminated dredgings from Rotterdam inner harbour and Hamburg are no longer dumped at sea. If the maximum estimates of metal inputs are accepted, atmospheric inputs are particularly important (Fig. 9.2(d)), but if minimum values are used, inputs from the atmosphere,

rivers, and dredging spoils are all of the same order (Fig. 9.2(e)).

The input of organochlorine pesticides and PCBs is small, following the phasing out of pesticides and control of the use of PCBs. However, DDT and 'drins' continue to be leached from agricultural land where they were formerly used and perhaps 200 kg yr^{-1} enter the North Sea via rivers. Rivers are also responsible for an input of 3 t yr^{-1} each of HCH and PCBs.

Atmospheric inputs, although large (see Table 5.3, p. 65) are dispersed throughout the whole of the North Sea. The effect of river inputs is most evident in the eastern North Sea and German Bight. Unlike inflows from British rivers which disperse rapidly, water from the continental rivers remains in a coastal strip as it flows northward up the Dutch, German, and Danish coasts. This is reflected in the higher metal concentrations in continental coastal waters than in British waters (Table 9.1). Mercury and lead tend to become associated with particulate matter rather than remain in solution; they are in low concentration in the water but contaminate coastal bottom sediments. The central North Sea is relatively isolated from these inputs, though it has a higher metal concentration than open ocean water. The continental coastal waters also receive a large river input of plant nutrients and, since this is an area where a summer thermocline often develops, are subject to phytoplankton blooms and oxygen depletion of the bottom water (see Fig. 2.15, p. 25).

Oil inputs to the North Sea are estimated to be between 71 000 and 150 000 t yr^{-1}, the chief uncertainties being the size of atmospheric and river inputs. The North Sea has had its share of oil tanker accidents, resulting in releases of 1000–5000 t in the worse cases, though since the Straits of Dover are too shallow to allow the passage of a fully laden VLCC, it has escaped catastrophic spillages on the scale of the *Amoco Cadiz* (see p. 29). The worst spillage was in the 1977 Ekofisk blow-out (see p. 32) when 20 000–30 000 t of crude oil were discharged, although that had

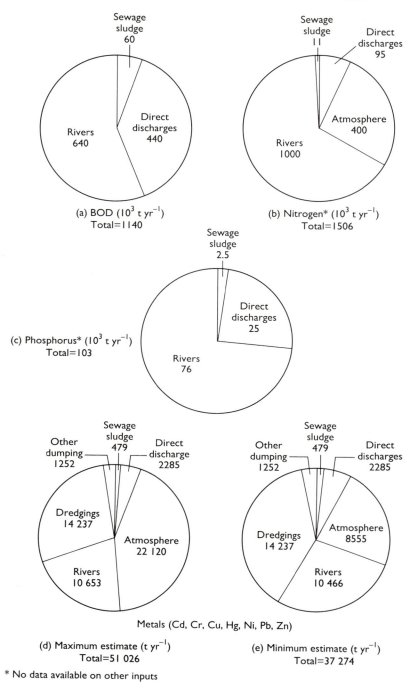

Fig. 9.2 Estimated inputs to the North Sea by various pathways.

* No data available on other inputs

no detectable effect. Discharges of oil production water from offshore platforms and the disposal of drill cuttings contaminated with oil-based drill muds account for 29 000 t yr^{-1}, but their effect is very localized.

As the North Sea has been designated a 'Special Area', it is now illegal for ships to

Table 9.1 Typical dissolved concentrations (μg l^{-1}) of trace metals in North Sea areas

Metal	Netherlands coast	German Bight	UK coast	Central North Sea	Average ocean
Cadmium	0.10	0.05	0.03	0.02	0.01
Copper	0.60	0.60	0.30	0.20	0.10
Zinc	5.0	2.0	4.0	0.6	0.1
Nickel	1.0	0.8	0.7	0.3	0.2
Mercury	0.01	0.01	0.01	0.001	0.001
Lead	0.05	0.05	0.05	0.03	0.003

discharge oily ballast and bilge water, but these practices continue and are responsible for the large number of small oil slicks that are reported in the North Sea. It is impossible to estimate the total amount of oil discharged in this way.

Impact on fisheries

Since the resumption of fishing in the North Sea after the end of the Second World War, there was a steady increase in the annual catch of most species. In 1971 the total catch was almost 3 m t; but in 1984 it was 2.4 m t. The fishing is monitored and regulated, and catch quotas are agreed each year, depending on the strength of recruitment to the fish stocks, but these quotas are often exceeded. Over-fishing has depressed the catch of some species, but a more important development has been the introduction of 'industrial' fishing of small fish for conversion to fish meal. Since the 1960s, Norway pout and sand eels, which were not previously fished, have formed the basis of industrial fishing, and these catches partly accounted for the continuing increase in the total North Sea catch. Young herrings were also taken for fishmeal after the early 1950s and this over-exploitation led to a serious decline in the herring fishery in the 1960s and contributed to its collapse at the end of the 1970s.

Although the strength of recruitment varies from year to year, the practices of the fishing industry have a dominating influence on the size of exploitable fish stocks in the North Sea. Against this background, it is impossible to detect any effects that pollution may have on the fishery.

There is some concern about the effects of sand and gravel extraction in the North Sea, since gravel beds are the spawning grounds of herring, and sand eels live in them permanently. Some 30–5 m t yr^{-1} of sand and gravel are dredged from European waters, much of it from the North Sea. Exhaustion of gravel beds on land and the projected needs of the construction industry suggest that the rate of exploitation of marine deposits will continue to increase. Besides destroying a vital habitat, this dredging may have indirect effects on fisheries. Ground containing as little as 30 per cent gravel can be mined and the fine material which is washed out increases the turbidity of the water and causes siltation, which may blanket the bottom fauna as well as release toxic materials from the sediments. Recovery of the fauna in the dredged areas is slow; one study off the south coast of England showed only slight recovery with a few polychaete species a year after dredging.

Because of the possible threat to human health and the known input of metals to the North Sea, contamination of food fish has been closely monitored. In most places, mercury concentrations in the muscle of cod and plaice is below 0.1 μg g^{-1} (p.p.m.), zinc contamination averages 5.2 μg g^{-1}, copper 0.65 μg g^{-1}, and cadmium 0.02–0.7 μg g^{-1}. These are all very low values and there is no evidence of serious contamination. The highest values occur in fish caught off the coasts of northern France, Belgium, and

Holland, and in the Thames estuary, reflecting industrial inputs. Herring contain less mercury (0.06 μg g^{-1}) than the demersal species, but slightly higher concentrations of copper and zinc. Because of their high fat content, herring contain greater amounts of organochlorine pesticides and PCBs than occur in cod or plaice, with the PCB content 2–5 times greater than pesticides; but the content varies widely in different samples and is nowhere very high. As might be expected, mussels acquire greater concentrations of these conservative materials then fish do—generally about three times greater. These, too, reflect local inputs, but commercial harvesting or culture of mussels and other bivalves takes place in uncontaminated areas.

Impact on sea birds

In addition to a large resident bird population, the North Sea and its coasts provide wintering grounds for very large numbers of shore and sea birds. The dense flocks of birds are vulnerable to oil pollution and there are numerous casualties each winter. The worst case was in 1955 when the tanker *Gerd Maersk* grounded near the mouth of the river Elbe and spilled 9000 t of crude oil. Estimates of the numbers of birds killed ranged from 50 000 to 500 000. In February 1969, a few hundred tonnes of fuel oil drifted among the Dutch islands and through the Waddensee, killing 35 000–41 000 birds, including a large part of the breeding population of eider. Small quantities of oil may also cause heavy casualties—12 000 on the east coast of Scotland in 1971 and 30 000 in the Skagerrak in 1981—but regular counts of dead oiled birds on North Sea coasts indicate that casual, small-scale oil pollution is a continuing hazard to birds in the North Sea. The greatest number of casualties is in winter, and the prevailing westerly winds drift both floating oil and birds towards the continental coast, where 70–80 per cent of dead birds on the beaches have been oiled (Fig. 3.10, p. 43). Despite these great and continuing losses, earlier fears that they were causing a reduction in the breeding populations of a number of sea ducks (scoter, long-tailed ducks, and so on) and auks, now appear to be unfounded. Sea bird colonies around the northern North Sea, in fact, appear to be stable or to have increased in size.

More insidious threats to the wintering bird flocks are the reclamation of coastal wetlands and shallows, destroying irreplaceable feeding grounds, and the depletion of fish stocks upon which the birds depend for food. It is not possible to say whether these activities have had any impact on bird populations so far, but the destruction of their food resources cannot be expected to continue without some adverse consequences.

International action

Atmospheric and river-borne inputs account for a considerable fraction of the wastes entering the North Sea, and these are derived from much of central and western Europe, not simply the population of the coastal zone. The river Elbe carries wastes from Czechoslovakia and all of Germany into the North Sea; the Rhine receives inputs from Switzerland, France, Germany, Luxemburg, and Holland; the Scheldt arises in Belgium, but flows into the North Sea through Holland. While it is within the power of individual states to improve the quality of some of their waters, in many cases, particularly in areas influenced by the outflow from continental rivers, only international action can achieve any improvement.

In 1974, under an international convention, the Oslo Commission was established to control discharges and dumping from ships into the North Sea. The comparable Paris Commission was set up a little later to regulate direct discharges from land. In addition, since most of the North Sea states are members of the EC, the European Commision has some power to regulate waste inputs and has acted, for example, by stipulating maximum permissible contamination of shellfish for human consumption and bacterial contamination of bathing beaches.

The International Council for the Exploration of the Sea (ICES) includes representatives of all the countries fishing the North Sea and since the early years of this century has been responsible for evaluating fish stocks and recommending annual fishing limits. For the last 20 years, it has also had an important role in arranging monitoring programmes and collecting together scientific information about pollution levels and impacts.

In recent years there has been increasing disagreement about the impact of wastes discharged into the North Sea; to agree future action, the Environmental Ministers of the North Sea states met in 1984, 1987, and 1990, and intend to meet regularly in the future. As a result of these meetings, there is now agreement about the extent and source of the more important wastes reaching the sea, and for phased reductions of inputs, particularly those affecting the most vulnerable areas in the eastern North Sea.

THE MEDITERRANEAN SEA

The Mediterranean is a deep, virtually tideless sea with a surface area of 2 965 000 km^3. Much of it is more than 200 m deep, with a number of deep basins below 3000 m. The eastern and western Mediterranean, separated by the relatively shallow straits between Sicily and Tunisia, show differences in resident fauna and flora, indicating a degree of isolation between the two regions. The Aegean, and particularly the Adriatic, are semi-enclosed extensions from the main body of the Mediterranean.

Evaporation is about three times greater than the input from precipitation; the deficit is made good largely by an inflow of Atlantic water at Gibraltar. There is negligible input to the eastern Mediterranean and, as a result, the salinity in the west is near oceanic levels, approaching 37 per mille, but near the coast of Asia Minor it is 39 per mille. This dense saline water sinks to a depth of 100–200 m, flows westward, and leaves the Mediterranean as a deep current at Gibraltar. Through most of the year there is good vertical mixing of the top 200 m, and sometimes 600 m, of water.

The main centres of coastal population are on the northern side of the western Mediterranean and around the head of the Adriatic (Fig. 9.3). These are also the most industrialized areas and receive the majority of the 100 million tourists that inflate the summer population of Mediterranean countries. The North African coast, by contrast, is for the most part arid with little urbanization or industrialization. Pressures on the marine environment therefore vary widely depending on the local and human environment.

Fisheries

The Mediterranean is only moderately productive and the demand for fish in the Mediterranean countries substantially exceeds the local supply. The deficit is only partly compensated for by imports from the Atlantic and other fisheries, and the cash value of the Mediterranean fisheries is therefore high, with prices five to seven times above average world prices. Bottom-living (demersal) species such as mullet, hake, and so on are most in demand. The main stocks of these are along the northern coast and these are fully exploited. A number of pelagic fish (for example anchovy, sardines, and mackerel) in the northwestern Mediterranean are also intensively fished. Mariculture is practised in some places, but is not widespread.

Oil pollution

All parts of the Mediterranean are chronically polluted with tar balls; a high value of 500 l km^{-2} has been recorded south of Italy, and many tourist beaches are irritatingly contaminated with specks of tarry oil. The main sources of oil pollution are deballasting and tank-washing operations of oil tankers and the discharge of oily bilge water by other shipping. Refinery wastes are a secondary, but important, source of chronic oil pollution. Offshore oil extraction and tanker accidents have been of minor significance. Some 250 m t yr^{-1} of oil are transported through the Mediterranean, about 150 m t yr^{-1} of this from North Africa to European ports (Fig. 9.4). The cross-Mediterranean journey time

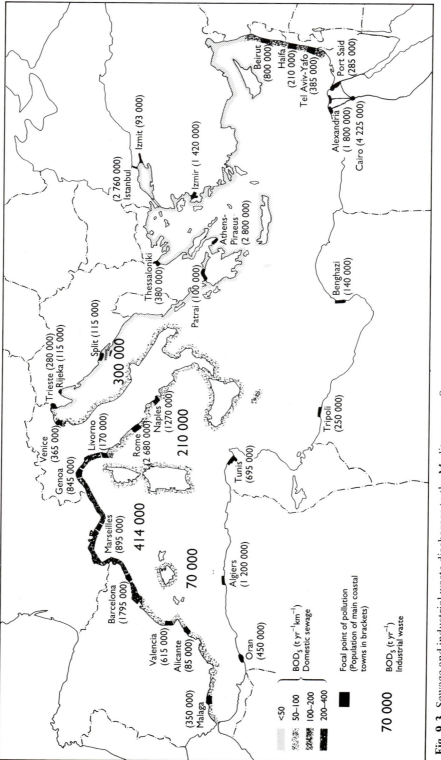

Fig. 9.3 Sewage and industrial waste discharges to the Mediterranean Sea.

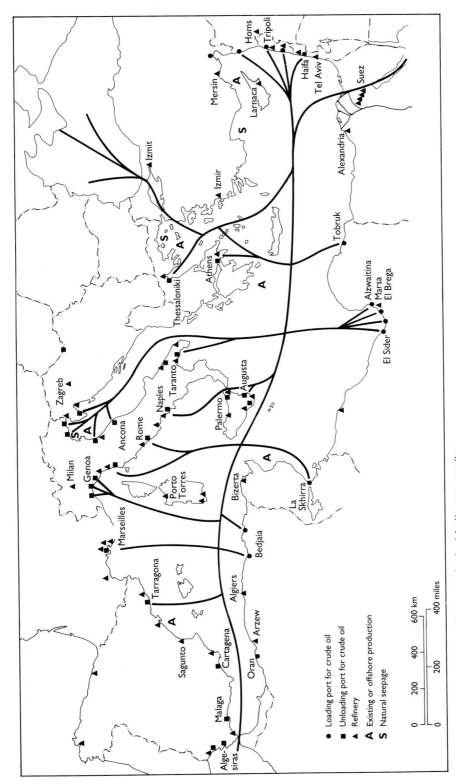

Fig. 9.4 Activities of the oil industry in the Mediterranean Sea.

is too short to allow efficient operation of the load-on-top system (see p. 28), and there have been inadequate slop-reception facilities in ports. A ban on the discharge of oily wastes in 1976 and stricter enforcement has proved effective and the quantity of floating tar in the eastern Mediterranean has declined from 37 000 μg m^{-2} in 1969 to 1175 μg m^{-2} in 1987.

As a result of oil pollution, tainting of a variety of fish and bivalves, rendering them unmarketable, has been reported from the neighbourhoods of oil ports in Spain, France, Italy, and Yugoslavia. Spiny lobsters have been killed by oil pollution around Bizerte in Tunisia. The Bay of Izmir and the Sea of Marmora, on the Turkish coast, have suffered adverse effects from oil and the spawning grounds of bonito and mackerel have been damaged. The most serious effects of oil pollution have been noted in the Gulf of Naples, Cagliari, and the Venetian lagoons, where fish populations have been reduced. The Bay of Muggia at Trieste, once rich in fish and molluscs is now described as being biologically almost a desert because of the impact of petrochemical wastes.

Domestic wastes

A survey made in 1972 found that most coastal communities discharged untreated sewage to the sea and the rest usually gave only minimal treatment. The worst affected areas were the riviera coast from the river Elbo in Spain, across the French riviera, to the river Arno in Italy (Fig. 9.3). It was estimated that the discharge amounted to 336 t km^{-1} yr^{-1}. The Israeli coast was found to be similarly affected. A vigorous attack on this problem was mounted, especially in France and Israel and on parts of the Spanish and Italian coast, but it is doubtful if the situation has much improved elsewhere. One difficulty is the need to provide sufficient sewage treatment facilities to cope with the enormous influx of population for a few summer months.

Apart from damage to the tourist industry for aesthetic reasons, discharges of untreated

sewage are a threat to public health. In 1973, there was an epidemic of cholera centred around Naples, spread by the consumption of infected shellfish, and lesser gastric disorders have been a common hazard to visitors to Mediterranean coasts for many years.

Other pollutants

It is estimated that 85 per cent of anthropogenic inputs to the Mediterranean are from the land, and inputs from rivers dominate. The organic load (BOD) from rivers equals that from coastal towns and most of the phosphorus and nitrogen inputs are from rivers. The input of 90 t yr^{-1} of organochlorine pesticides is almost entirely derived from the drainage areas of rivers. The most considerable input of heavy metals and PCBs is from the rivers Rhone and Po, added to those from the heavily industrialized zones near their mouths (at Marseille and Toulon and at the head of the Adriatic, respectively).

The Gulf of Fos influenced by the Rhone and its industrial belt, and the upper Adriatic, together with the Saronikos Gulf receiving the effluents of Athens and Piraeus, have been studied in some detail and all show some impact of the effluent load they receive. There is most concern about the conditions of the Venetian lagoons and upper Adriatic which, because of the low rate of water exchange and their relative isolation from the rest of the Mediterranean, are least able to accept a heavy load of effluents without damaging consequences.

Phosphorus is the limiting nutrient for phytoplankton in the upper Adriatic and inputs of phosphorus to the area have increased markedly. The upper Adriatic now receives about 30 000 t yr^{-1} of phosphorus, of which 60 per cent is contributed by the river Po. Half the phosphorus is derived from detergents and agricultural fertilizers which came into intensive use after 1945. The progressive eutrophication resulting from this is shown by the increased oxygen concentration in surface waters and oxygen depletion of bottom waters, which first became apparent in 1955 around the mouth of the river Po but now includes much of the western and northern parts of the upper Adriatic (Fig. 2.16, p. 26).

Despite the existence of small areas where conservative materials accumulate in undesirably high concentrations, neither the water nor the organisms in the Mediterranean appear to be seriously contaminated and concentrations are generally comparable to those in the open Atlantic. In the early 1970s, a few fishing communities on the Italian Adriatic and French coasts were found to be exposed to high levels of mercury due to local inputs contaminating seafood organisms in the area. These inputs were subsequently brought under control. Mercury levels in two species of tuna have also been found to be high, at 2–3 p.p.m., three times greater than those in Atlantic tuna. This appears to be due to natural causes, perhaps from the greater level of seismic activity in the Mediterranean. Mussels from the coast near Marseille contain high concentrations of copper (95 p.p.m.), and high concentrations of zinc (200 p.p.m.) occur in coastal waters in a number of areas, but neither constitutes a health risk.

There is evidence that PCBs in the Mediterranean are subject to 'demagnification'. Microplankton strongly accumulate PCBs in areas with a large local input, but contamination of animals at higher trophic levels is progressively reduced. The faecal pellets contain ten times the concentration of PCBs found in the animals that produce them and when the pellets sink to the seabed in the deeper parts of the Mediterranean, they are removed from pelagic food chains. The same phenomenon has been reported for radionuclides.

International action

Many of the problems in the Mediterranean are local in character and are susceptible to local solutions. While the financially important fisheries have not been generally affected by pollution and fish catches continue to grow, the Mediterranean receives a very heavy effluent load which, because of the restricted water exchange, poses a continuing threat. The United Nations Environment Programme (UNEP) has provided a focal point for international cooperation in cleaning up the Mediterranean, and this culminated in the 1976 Barcelona Convention to put this into action.

Subsequent progress has been slow for a variety of reasons. There are 18 Mediterranean states (21 if Black Sea states are included) and it is difficult to achieve concerted action by so many countries. The problem is compounded by severe political obstacles in some cases, and by the great diversity in the degree of development of different coastal states, with corresponding differences in the priority accorded to environmental protection.

THE BALTIC SEA

The Baltic is the largest body of brackish water in the world, with an area of 370 000 km². It consists of several basins of various depths, the greatest being 459 m, separated from each other by relatively shallow sills (Fig. 9.5). Communication with the North Sea is through the narrow Öresund between Denmark and Sweden, which is only 7–8 m deep, and the Belt Sea between the Danish Islands of Fyn and Zealand, where the depth is 17–18 m.

The annual input of freshwater from the land varies from 440–70 km³, depending on climatic factors, and the inflow of salt water from the North Sea is about the same. These inputs are balanced by an equivalent outflow into the Kattegat. Surface salinity is below 5 per mille in the inner parts of the Gulfs of Bothnia, Finland, and Riga, 5–7 per mille in the main part of the Baltic, and 7–9 per mille around the south of Sweden. This brackish water floats on denser more saline deeper water and, while wind and thermal convection currents produce some vertical mixing in autumn and winter, there is a permanent halocline at a depth of about 40 m in the southwestern basins, and at 60 m in the central basins, separating the surface layer from the deeper water (Fig. 9.6).

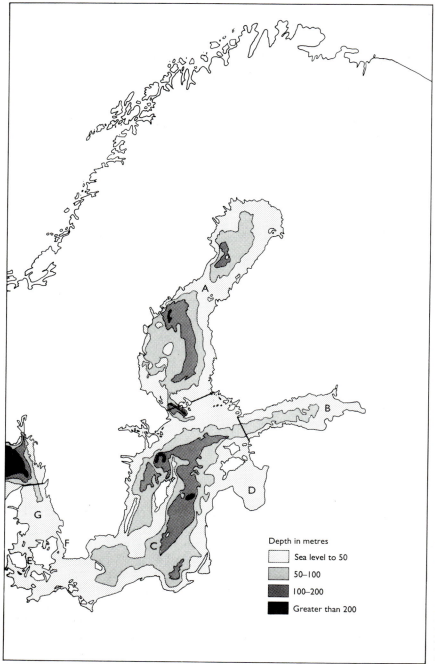

Depth in metres

	Sea level to 50
	50–100
	100–200
	Greater than 200

Fig. 9.5 Deep basins of the Baltic Sea, contours at 50 m, 100 m, and 200 m. A, Gulf of Bothnia; B, Gulf of Finland; C, Baltic proper; D, Gulf of Riga; E, Belt Sea; F, Öresund; G, Kattegat.

Brackish Baltic water flows out to the North Sea, generally over an inflowing deeper current of North Sea water. Depending on wind direction and speed, the flow over the sills in the Öresund and Belt Sea may be entirely of Baltic water flowing seaward, or of North Sea water flowing inward. The latter is important because it is only during strong

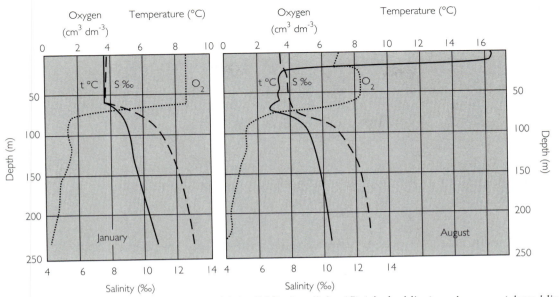

Fig. 9.6 Distribution of temperature (*t*) (*solid line*), salinity (*S*) (*dashed line*), and oxygen (*dotted line*) with depth in the Gotland Deep in January and August 1978. (*Pergamon Press.*)

inflows from the North Sea that water in the deepest parts of the basins is replaced. At such times a second halocline develops at a depth of 110–30 m. Once the deep basins have been recharged with salt water, the density of the new bottom water must be substantially reduced by mixing before a new influx can replace it. This may take five years, and during this period the bottom water is essentially stagnant.

During these periods of stagnation, the deep basins become depleted of oxygen through the bacterial degradation of organic material present in the sediment and drifting down from the surface layer. Extensive areas with anoxic conditions and the production of hydrogen sulphide therefore occur at irregular intervals (Fig. 9.7), and at such times the benthic fauna disappears until a new influx of North Sea water allows recolonization.

Organic wastes

In the 1960s when severe and increasing deoxygenation of bottom waters was

detected, there were fears that the Baltic had become catastrophically polluted and was dying.

The Baltic is naturally oligotrophic—it has a small natural organic input and low production—but receives a large quantity of organic wastes and shows some signs of eutrophication. A population of 17.5 million people living around the Baltic is responsible for the direct input of 2.3×10^6 m^3 day^{-1} of domestic sewage, of which 40 per cent is untreated. A further 1.2×10^6 m^3 day^{-1}, 20 per cent of it untreated, reaches the Baltic via rivers. Domestic sewage is the most significant organic input in the Gulf of Finland and the southwest Baltic, where there is the greatest density of population. In the Gulf of Bothnia the chief input of organic matter is from the paper and wood pulp industries. The principal inputs of nitrogen and phosphorus, in both inorganic and organic forms, are largely from agriculture in run-off to rivers, but on the south coast of the Gulf of Finland they also arise from several fertilizer factories.

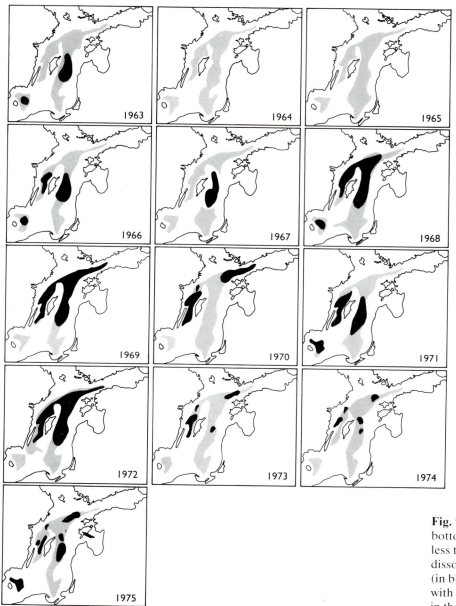

Fig. 9.7 Areas with bottom water containing less than 2 mg l^{-1} dissolved oxygen and (in black) anoxic areas with hydrogen sulphide, in the Baltic, 1963–75.

The installation of industrial and urban treatment plants in the 1970s reduced the input of phosphorus by 50 per cent but, despite the reduction of inputs from these sources, nitrogen from agricultural run-off increased and between 1969 and 1977 the net input of nitrogen to the Baltic increased by 10 per cent.

The effects of the increased input of organic matter and plant nutrients have been studied in coastal waters. In inshore waters, the primary production of phytoplankton has greatly increased, reducing the transparency of the water. On the shore, green algae and emergent plants in shallow water show enhanced growth. There is some evidence of an increase in zooplankton biomass throughout the Baltic, as a result of the phytoplankton increase, but in coastal waters this has been accompanied by a change in species composition of the zooplankton. The relative importance of predatory fish has decreased and although the total fish biomass has increased, catches in coastal waters, though high, are of less desirable species and so of reduced value. Less is known of the effect of the enrichment of offshore waters, though heavy phytoplankton blooms, particularly of blue–green algae, have increased and there are fears that the increased input of detritus to deep water that results from the blooms may accelerate the development of anoxic conditions.

To a degree, enrichment of the waters is beneficial to fisheries and the incursion of North Sea water allows the greater penetration of herring, sprat, and cod into the Baltic, although they do not live in the gulfs where salinity is below 5–6 per mille. These marine species make the most significant contribution to fish biomass in the Baltic and account for 90 per cent of the landings. The fishery for migratory eel and salmon is economically important, although the total weights landed are small. Freshwater commercial species, perch, pike, roach, and white fish (*Coregonus*) are caught in the waters of lowest salinity, but are of secondary importance. The overall yield of fish from the Baltic is fairly low, 2.2 g m^{-2}, compared with 5.2 g m^{-2} in the North Sea, and Baltic fish grow more slowly; but there is no evidence that the offshore fisheries have suffered from eutrophication of the waters. There is some sign, however, that sprat and herring stocks are over-exploited and may decline for that reason.

Conservative pollutants

In the 1960s, along with fears about the effects of eutrophication, there was great concern at high concentrations of mercury found in fish from the inland and coastal waters of Denmark, Finland, and Sweden; in Sweden, commercial fishing was banned in areas where the fish contained more than 1 μg g^{-1} of mercury. Fish in the open sea were not affected. The mercury was derived mainly from pesticides, particularly from fungicides and slimicides used in the paper and wood pulp industry in Finland and Sweden.

Following this, the use of mercurial pesticides was much reduced and the occurrence of heavy-metal contamination in sediments and a number of marine organisms has been closely monitored in most parts of the Baltic.

Much of the metal input (Table 9.2) is carried to the seabed by sedimentation, and recent benthic sediments contain 200 per cent more mercury and 100 per cent more cadmium than the natural background levels. Lead, zinc, and copper levels are also significantly higher. The most contaminated sediments are in the Kattegat, Öresund and Arkona basin, in the Gdansk Bight, and in the Gulf of Bothnia. Metal contamination of fish (Table 9.3) has now fallen to a more acceptable level and is not significantly higher than in fish from the north Atlantic. The highest concentrations, as might be expected, are found in the neighbourhood of the Öresund, Gdansk Bight, Gulf of Finland, and Gulf of Bothnia. For cadmium, in particular, contamination is higher in inshore animals than offshore.

Halogenated hydrocarbons, on the other hand, continue to cause damage. The use of DDT has generally ceased and residue levels in fish and guillemot eggs continued to fall

Table 9.2 Total metal inputs to the Baltic Sea (t yr^{-1})

Mercury	Cadmium	Lead	Nickel	Copper	Zinc
30*	250*	4300–6200*	2100–5100†	4500–8500*	15 500–18 900†

* 50–80 per cent atmospheric.
† 50–80 per cent from rivers.

Table 9.3 Heavy metals in marine organisms from the Baltic

	Mercury (p.p.b.)	Cadmium (p.p.b.)	Copper (p.p.m.)	Zinc (p.p.m.)
Cod	20–880	2–50	0.08–2.4	1.2–9.2
Herring	4–90	2–200	0.30–1.9	3.4–32
Flounder	10–450	2–100	0.10–0.89	3.5–11.3
Mussel	4–300	130–560	1.0–16	16–110

during the period 1974–82. Despite restrictions on their use, there are still land-based inputs of PCBs, particularly near Turku, in Finland, and Stockholm, and contamination of fish and guillemot eggs by these compounds has not shown a reduction since 1974. Apart from a few hot-spots, there appears to be no geographical gradient in the contamination of fish by DDT or PCBs and the level of contamination is not unduly high. The principal dangers of halogenated hydrocarbons are to fish-eating birds and mammals, and in the 1970s concentrations in seals and birds of prey were ten times higher in the Baltic than in areas west of Sweden. The sea eagle, *Haliaetus albicilla*, was threatened and the reduction of the breeding populations of ringed seal, *Phusa hispida*, and grey seal, *Halichoerus grypus*, in the Baltic (both of which are heavily contaminated with PCBs) still gives cause for concern (see p. 96).

International action

The Helsinki Convention, which came into force in 1980, is concerned with the protection of the Baltic Sea. All eight states with Baltic coastlines are active members of the Commission which puts the Convention into effect. Like the Oslo and Paris Commissions for the North Sea, the Helsinki Commission regulates discharges and dumping from ships and direct discharges from land. It also has wider responsibilities and has been concerned with such matters as atmospheric inputs, emergency action against oil spills, the use of antifouling paints containing tributyltin, co-ordinating monitoring programmes, and seal conservation. It frequently arranges meetings of experts to evaluate the scientific evidence on which its recommendations are based. The Helsinki Commission has proved to be very effective: through its efforts the Baltic is, with the North Sea, one of the most thoroughly studied seas in the world.

THE CARIBBEAN

The greater Caribbean area includes the Caribbean Sea and the Gulf of Mexico. Its total surface area is 4.24×10^6 km^2, and it consists of a number of deep basins separated by major sills. It lacks a shallow continental shelf, except in the Yucatan Gulf, off the US Gulf coast, and part of the South American shelf; 80 per cent of the water is deeper than 1800 m and half is deeper than 3600 m, with the greatest depth of 7100 m in the Cayman Trench.

The Caribbean Sea and southern half of the Gulf of Mexico are tropical and experience little seasonal change. The surface water temperature is 27 °C with a seasonal fluctuation of less than 3 °C. The northern part of the Gulf of Mexico, however, has a winter surface sea temperature of 16 °C, rising in summer to 28 °C. Over much of the whole area there is a permanent thermocline at a depth of about 100 m and upwelling is not a dominant feature, although there are some local areas where bottom-water comes to the surface.

Because of the existence of the thermocline and lack of upwelling, much of the area is deficient in nutrients, and significant fisheries, particularly for penaeid prawns, are confined to the shallows off the Mexican and US coasts. These are associated with the highly productive coastal ecosystems of mangroves, seagrass, and coral reefs. The US commercial fishery in the Gulf of Mexico is worth $400 m per year, and the income from recreational fishing is at least as great.

Seagrass beds (Fig. 9.8), principally of turtle grass, *Thalassia testudinum*, can contribute 3000 g m^{-2} yr^{-1} of primary production (dry weight), of which two-thirds is due to seagrass growth and one-third to the associated algae. This is the same as, or even greater than, the production of mangrove swamps. On coasts fringed by mangroves, with seagrass beds beyond them and coral reefs to the seaward side, very high levels of productivity are reached, with fish leaving the shelter of the reef to feed in the seagrass beds, and both seagrass and mangroves, besides harbouring a large and varied fauna, providing important nursery grounds for prawns and fish.

Another feature of mangroves and seagrass beds is that their root systems stabilize the seabed and prevent coastal erosion. They also clarify the water by trapping sediments.

In the Caribbean region as a whole, there is relatively little coastal urbanization or industrialization, and there is no widespread pollution except from oil. In a number of areas, however, there has been severe local impact from a variety of pollutants (Table 9.4).

Oil pollution

The principal oil-producing areas are Trinidad and Tobago, Venezuela, and the Gulf of Mexico. Nearly one-third of the production is from offshore oilfields and there are more than 2000 fixed offshore platforms in the US sector of the Gulf alone. Blow-outs, overflows, pipeline fractures, and other accidents at the platforms are the major source of oil pollution in the area.

Studies of the benthic fauna in oilfields off the Louisiana coast have failed to reveal any significant differences from other areas, and the platforms themselves, providing the only hard structures in a generally sedimentary region, attract an exotic attached fauna and flora, as well as sheltering a variety of sport fish. Despite the spillages and some contamination of sediments, offshore activities appear not to have a serious environmental impact.

The largest oil spill in history was that following the blow-out in the Ixtoc I field in the Bay of Campeche off the Mexican coast, from June 1979 to March 1980, when the well was finally capped. Altogether 350 000 t of crude oil were released into the sea. Prevailing offshore winds prevented fresh oil coming ashore on the Mexican coast, but 80 km of the Texas coastline, some 800 km away, was seriously affected and clean-up costs there

Table 9.4 Incidence of severe local pollution damage in the Greater Caribbean

Location	Pollutant
Cartagena, Colombia	Mercury
Colon, Panama	Oil
Veracruz, Mexico	Thermal effluent
Galveston, Texas	Oil
New Orleans, Louisiana	Industrial waste
Tampa Bay, Florida	Sewage
Kingston, Jamaica	Bauxite waste
Port-au-Prince, Haiti	Sewage
Santo Domingo, Dominican Republic	Urban run-off
San Juan, Puerto Rico	Sewage
Port of Spain, Trinidad and Tobago	Oil
Cumana, Venezuela	Sewage
Havana Bay, Cuba	All the above

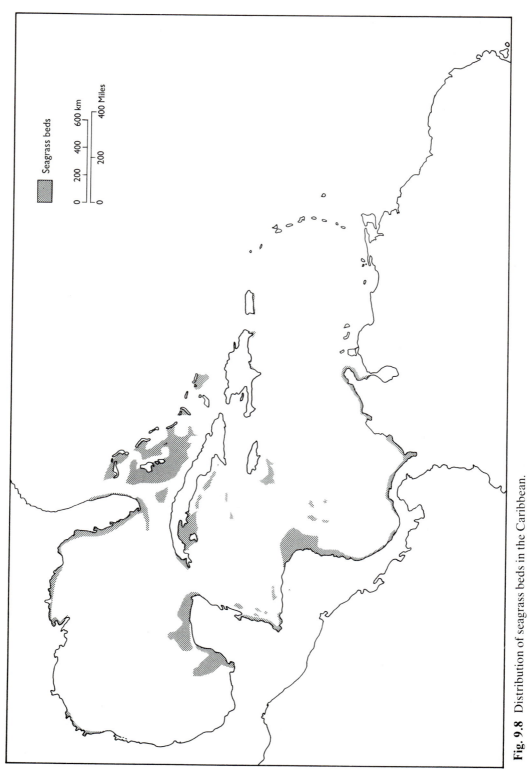

Fig. 9.8 Distribution of seagrass beds in the Caribbean.

exceeded $4 m. Entrances to the biologically sensitive Laguna Madre were boomed off to protect the breeding grounds of the endangered brown pelican, *Pelicanus occidentalis*; 10 000 baby Kemp's Ridley turtle, *Lepidochelys kempii*, another endangered species, were airlifted from the only breeding ground of this species on beaches on the north coast of Mexico to another area of the Gulf of Mexico which was free from oil. The total cost of capping the well and containing the environmental damage, including lost oil revenues, amounted of $219 m but, despite the enormous quantity of oil spilled, the environmental consequences were remarkably small. The $230 m per year Mexican and US shrimp fishery in the Gulf escaped long-term damage.

Heated effluents

The sea temperature is close to the upper thermal limit of many organisms and they are therefore vulnerable to thermal pollution. This is not a widespread hazard because of the low level of industrialization of coasts in the Caribbean, but cooling water discharges from two power stations in Puerto Rico have caused extensive damage to seagrass beds in Guayanilla Bay. Plants are killed and production is reduced in areas where the sea temperature is 5 °C above the ambient temperature of 30 °C. The benthic fauna in shallow waters is completely eliminated at this or higher temperatures.

Sedimentation

Turbid water reduces the productivity of the seagrass beds and heavy sedimentation is lethal to corals. The Mississippi and a number of lesser rivers flow into the Gulf and Caribbean and most of the water from the Orinoco travels north and enters the southern Caribbean. These rivers carry a heavy sediment load, particularly from areas where bad farming practices result in soil erosion. The damaging effects of these inputs is limited to the areas, sometimes large, around the mouths of the rivers.

More widespread damage is caused by coastal developments involving dredge and fill, or the dredging of shipping channels. Dredging operations cause increased turbidity of the water and marine dumping of the spoil extends the damage over a wider area.

The common practice of extracting beach sand for construction purposes, as well as coastal developments, largely for the tourist trade, leads to destruction of mangrove forests and seagrass beds and has caused severe coastal erosion in a number of places. This, in turn, leads to increased turbidity of the water and increased sedimentation elsewhere.

Tourism

The Greater Caribbean receives about 100 million tourists annually and tourism represents a major industry, particularly in the eastern islands (Fig. 9.9). The attractions are sun, white sandy beaches, and warm clear waters. Hotel development is almost entirely on the coasts and often in isolated places. There is a general lack of awareness about the environmental impact of such developments, and while some hotels install their own sewage treatment plants, many do not or have inadequate treatment facilities. If, as is commonly the case, the coastal developments have resulted in the loss of seagrass beds and mangroves, leading to increased turbidity of the water, the combination with sewage pollution is detrimental to the interest of the tourist industry, even though the effects are quite local.

International action

Under the Regional Seas Programme of the United Nations Environment Programme (UNEP), an Action Plan for the greater Caribbean was adopted by 22 states of the region in 1981. UNEP provides the co-ordinating secretariat for the plan. Since then, a survey has been made of the special needs and problems of the Caribbean, and in 1983 a Convention was adopted by the major coastal states which requires the signatories to: prevent, reduce, and control pollution; safeguard sensitive environments; and collaborate

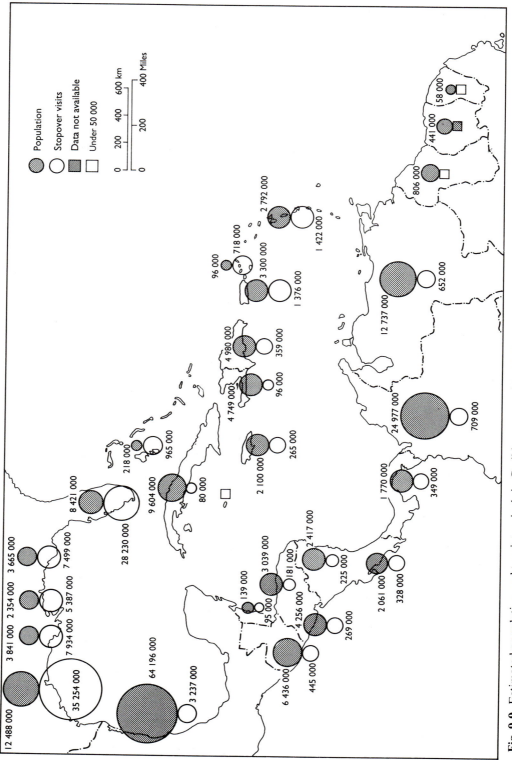

Fig. 9.9 Estimated population and tourist arrivals in Caribbean countries, 1977.

in scientific research and monitoring. Particular emphasis is placed on oil pollution.

THE CASPIAN SEA

The Caspian is by far the world's largest land-locked body of water, with a total area of 436 000 km². The northern part is very shallow, only 10–12 m deep, but the central Caspian has depths to 700 m and the southern part, separated from it by shoals in the region of the Apserhon peninsula, has depths to 1000 m.

The surface sea temperature is 24–7 °C in summer, but falls in winter to 9 °C in the south and 0 °C in the north, where extensive sea ice is formed. Rivers flowing into the Caspian add 350 km³ yr⁻¹ of freshwater, three-quarters of it from the Volga, but an amount equivalent to this plus the 100 km³ of precipitation directly on to the sea, is lost in summer by evaporation. There is little variation of salinity with depth, but it differs widely in different parts of the Caspian. In the northern shallows it is 5–10 per mille, falling to 2 per mille near the Volga delta, and in the central and southern parts it is 12–13 per mille. The eastern side of the sea receives little precipitation and has negligible river input, and the rate of evaporation is high. Bays and semi-enclosed areas of the sea here have very high salinities, reaching 200 per mille in the Kara-Bugaz Gulf which is almost completely separated from the Caspian proper. The composition of the salt differs from that of open ocean water, with much more sulphate, magnesium, and calcium, and less sodium and chloride.

There is little vertical circulation of water and this leads to serious depletion of oxyen below 200–300 m, and zero concentrations near the bottom, resulting in the production of hydrogen sulphide in the deeper sediments. The benthic fauna of the central and southern parts of the Caspian is therefore impoverished and, in parts, totally absent (Fig. 9.10). That of the northern Caspian is dominated by molluscs and previously was conspicuously lacking in polychaetes. Molluscs have a lower

nutritive value than have worms to fish, and growth rates of commercial fish species such as bream and sturgeon were lower here than in the productive Sea of Azov, which has a typical benthic fauna with abundant worms and crustaceans. In 1939–41, 65 000 *Nereis diversicolor* were introduced in an attempt to restructure the benthic ecosystem. By 1948 this worm had become established over 30 000 km², especially in the northwestern area of the sea where densities of 1800 m⁻², representing a biomass of 86 g m⁻², occurred. Molluscs still dominate the benthos but, with the availability of polychaetes, sturgeon now feed almost exclusively upon them instead of (as formerly and still in other parts of the Caspian) on chironomid larvae, crustaceans, fish, and molluscs. *Nereis* now forms 50 per cent of the diet of the bream.

The Caspian is subject to wide fluctuations in sea level and salinity, depending upon river inputs of freshwater. Prolonged drought in the Volga basin during the 1930s caused a fall in level of 2 m and a corresponding increase in salinity, leading to a crash in the average biomass of the benthic fauna in the northern Caspian from 36.84 g m⁻² in 1935 to 4.74 g m⁻² in 1938. There was a subsequent recovery in the early 1940s, but dams constructed on the Volga, Terek, and Kura rivers for hydroelectric power and irrigation schemes have again reduced the flow of the rivers into the Caspian Sea.

Pollution

Throughout the 1960s there was great concern about the effects of lowered sea level and increasing salinity, coupled with severe pollution on the important Caspian fisheries.

A principal source of pollution in the northern Caspian has been the Volga itself, with untreated sewage discharged from 15 towns on the river and its tributary, the Kama, and a large number of industrial plants also discharging their wastes.

The Azerbaijan coast is the site of a major oil industry, with numerous offshore oil wells, refineries, and petrochemical plants on the coast, and a large urban population around

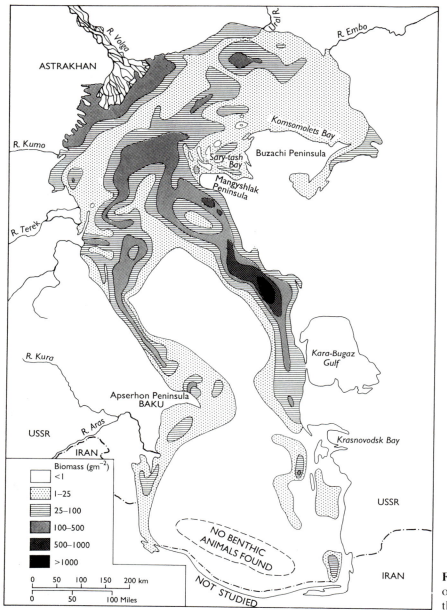

Fig. 9.10 Distribution of benthic biomass in the Caspian Sea.

Baku. Together they have been responsible for very damaging pollution of the previously productive western parts of the central and southern Caspian Sea. In the mid-1960s it was estimated that 1 m t of oil and petroleum products were released into the sea from accidental spillages, seepages, pipeline fractures, industrial effluents, and in refinery waste water. Added to this, the Caspian Steamship Line, the Caspian tanker fleet, and fishing vessels discharged tank washings, oily bilge water, and fuel oil to the sea. Sewage from Baku received only mechanical treatment before discharge, 36 sewer outfalls discharged untreated sewage, and 23 000 m^3 day^{-1} of untreated industrial effluent were discharged into the nearby Bakinskaya inlet.

These discharges caused severe contamination of the water and seabed with petroleum hydrocarbons: 27 mg l^{-1} around Sumgait, 46 mg l^{-1} in the Neftyanye Kamni region, and 148 mg l^{-1} in the Bakinskaya inlet. Phytoplankton diversity in the western Caspian fell from 74 to 40 species and biomass from 8.7 g m^{-2} to 2.1 g m^{-2}. The biomass of benthic organisms in coastal areas fell from 1724 g m^{-2} in 1961 to 21 g m^{-2} in 1969; the principal decline was observed in breeding and nursery grounds for sturgeon, bream, pike, crucian carp, and other commercial fish species.

For a variety of reasons—fall in sea level, increased salinity, and pollution (and overfishing cannot be excluded)—fish catches dropped dramatically. In the mid-1930s the total catch from the Caspian was 300 m kg (herring were not then taken); by 1970 the total catch was 110 m kg, of which 100 m kg were herring. Catches of sturgeon in the Azerbaijan region fell from 2100 kg yr^{-1} in 1931–5 to 350 kg yr^{-1} in 1961–4.

In 1972, a series of measures was announced to provide sewage treatment for towns on the Volga, to treat industrial wastes, and to reduce oil pollution in the Azerbaijan region of the Caspian. These measures have not proved effective, however, and waste discharges and contamination levels in the southern Caspian remain high. Fish catches have continued to decline and by the mid-1980s salmon and shad were reported to have almost completely disappeared and catches of bream and carp had fallen by a further 80 per cent; low value herring now comprise four-fifths of the total catch. Experience since the 1950s has demonstrated the difficulty of carrying on a valuable fishery together with rapid industrialization, massive irrigation schemes, and exploitation of offshore oilfields in a sea subject to the natural pressures of high evaporation and a variable river input of freshwater.

10

ASSESSING POLLUTION DAMAGE

THE NEED FOR QUANTITATIVE ASSESSMENT

Since pollution is by definition damaging, the obvious remedy seems to be to stop discharging the polluting substances into the sea. In recent years a number of measures have been introduced to do this but, unfortunately, pollution problems are not so easily solved. Different sectors of the environment (land, sea, and air) are not isolated from one another and a significant fraction of the wastes entering the sea are derived from rivers or the atmosphere. This is not a reason for ignoring direct inputs to the sea or for failing to control them, but much wider issues must also be taken into account. Waste disposal must be managed in such a way as to reduce environmental damage in all environmental sectors, not simply the sea. This demands careful assessment of the impact of a particular discharge. Both quantifying and assessing an impact introduce technical problems and call for clear thinking about what is meant by pollution.

Practicality

It has to be accepted that, for purely practical reasons, it is impossible to eliminate pollution completely.

The distressing loss of sea birds due to oil pollution is due almost entirely to human error or equipment failure leading to accidental spillages or to deliberate, but usually illegal, discharges of waste oil or oily water at sea. It is not difficult to devise safeguards to reduce oil pollution, but the only way to prevent all losses of sea birds caused by this source is to stop transporting or using oil at sea, which would eliminate the use of oil and oil products in many countries and effectively ground all aircraft. Most people would regard this as too high a price to pay: some sea bird mortality must be accepted as a consequence of using oil.

This may appear an extreme example of the impracticability of abolishing a form of pollution, but a similar argument can be used for many other pollutants.

Replacement therapy

While it is impracticable to eliminate oil pollution by ceasing to use oil, some other substances which are damaging to the environment can be withdrawn from use and replaced by less damaging substances.

This is an attractive solution to pollution problems, but care has to be taken to ensure that the replacement is not itself damaging. These problems are not easily solved, as the following examples show.

1. Mercurial slimicides used by the wood pulp industry were replaced by pentachlorophenol to reduce mercury releases to the sea. Pentachlorophenol has now been proved to accumulate in at least one organism (p. 68). Substances which have the potential to accumulate in marine organisms have a high hazard rating (see below).

2. Copper is very toxic to a wide spectrum of marine organisms and, because of this, was

valuable for use in antifouling paints on ships. However, copper leaching from antifouling paints caused damage in areas with a high density of shipping. Copper was replaced by tributyltin, which has proved to be even more damaging, particularly to commercial shellfisheries. Its use is now much reduced.

3. Organochlorine pesticides had a severely damaging effect on populations of predatory birds. DDT and the drins were therefore withdrawn from use in Europe and North America and the bird populations recovered. Organochlorine pesticides have low toxicity to humans and are safe to use. Their replacement by organophosphorus insecticides, which require special precautions in their use, continues to result in human deaths (p. 99).

4. Lead is seriously damaging to human health on land and so its use is progressively reduced. Lead pigments were formerly widely used in white paints and titanium dioxide, which has low toxicity, replaced lead for this purpose. Solving the public health problem created a different environmental problem in the sea, caused by the discharge of acid iron wastes (p. 79).

Precautionary principle

Fears about the damaging effects of wastes entering the North Sea led to the development of the **precautionary principle** which was proposed by Germany in 1986. It argues that, since we cannot reliably predict the effect of new inputs added to an area that already receives a large volume of wastes, no wastes should be discharged to the sea unless they can be shown to be harmless. This is a very stringent requirement because it is almost impossible to prove that a waste is harmless and, indeed, there are very few wastes that will not cause some environmental adjustment where they are discharged.

While the principle has received general acceptance, at least in a somewhat less stringent form, it does not address the fact that river and atmospheric inputs to the North Sea are often more significant than direct dis-

charges. Furthermore, it is concerned solely with protecting the sea and does not take into account the environmental consequences of disposing of the wastes in a different environment. One major criticism of the way in which the principle has sometimes been applied is that the slightest suspicion that a waste had a damaging effect was accepted as sufficient to warrant its control, without any consideration of scientific evidence. If the principle was strictly applied, there would be no place for science in decision-making.

Environmental impact assessment

In most developed countries, waste discharge requires some kind of official consent or licence. When a new discharge is proposed, the discharger must provide a detailed **Environmental Impact Assessment** (EIA) or **Environmental Impact Statement** (EIS) which predicts the effect of the proposed discharge on the surrounding area. On the basis of this, a permit may be issued or denied, or modifications of the discharge be required. The requirement to provide EIAs has been in force in North America for a number of years. In countries where a formal assessment is not required, very much the same inquiry is made before a permit to discharge a waste to sea is granted.

A recent analysis of the accuracy of the predictions made in EIAs submitted in Australia has shown that slightly more than fifty per cent of the assessments underestimated or overestimated the environmental impact of the developments. There have been no other surveys of the reliability of EIAs, but it is unlikely that the experience in Australia is unique.

Best available technology

Another approach to reducing environmental damage in the sea is to insist that waste discharges should be treated by the **Best Available Technology** (BAT) to minimize the release of damaging substances. This may seem a logical principle, but it has proved controversial.

The use of the best available technology may be over-protective if a small quantity of non-persistent waste is discharged into a large body of water, or it may give inadequate protection if several different sources discharge into a limited environment; the technology used to remove the noxious constituents of the waste may result in damaging consequences in a different environment; and, finally, a strict insistence on the use of the best available technology pays no regard to the cost of the treatment.

The high, and sometimes unnecessary, cost of using the best available technology has led to some resistance to its widespread introduction. As a compromise, the **Best Available Technology Not Entailing Excessive Costs** (BATNEEC) or the **Best Practicable Technology** (BPT) have been widely accepted, although they offer a weaker protection of the marine environment.

Best practicable environmental option

Industrial societies inevitably produce a great variety of wastes in large quantities. By fostering industrial processes that produce little or no waste (**low waste and no waste technology**) and by withdrawing particularly damaging substances (such as certain pesticides) from use, it is possible to reduce pollution. Even so, large quantities of waste will always be produced (for example sewage) and have to be disposed of somewhere; if not in the sea then in another environment. Waste disposal always has some impact on the environment, and in dealing with the pollution problems in one environment—the sea—it is only too easy to suggest solutions which merely transfer the problem to a different environment.

In countries like Britain or the Netherlands, which are densely populated and where land is relatively scarce, it is important not to solve a problem in the sea by creating a greater one on land. What is sought is the means of disposing of a waste that will cause least environmental damage. This is known as the **Best Practicable Environmental Option**. It may mean that in some instances disposal of a waste to sea, even though it causes some

damage to marine resources, is preferable to any other method of disposal.

Cost

The principle on which pollution abatement costs are allocated in most countries is that **the polluter pays**. It seems only just that those responsible for processes that could potentially pollute the environment should bear the cost of preventing this. In other words, the cost of a product or an activity should include the associated costs of environmental protection.

However, the polluter often turns out to be the community at large. Urban sewage represents by far the greatest volume of organic waste entering the sea and is responsible for the anoxic state of industrialized estuaries; one-third of the mercury entering the sea as a result of human activities comes from burning coal; and half the lead comes from the use of leaded petrol.

We have already seen that the cost of pollution abatement can rise very steeply if high environmental standards are required (see Fig. 1.1, p. 8). For most industrial processes, it is very costly to fit pollution control equipment or to modify existing industrial processes to meet new, more stringent environmental standards. The manufacturer has no alternative but to pass these costs on to the consumer.

The cost of reducing pollution from most sources falls on the community at large, directly through taxes or indirectly through other costs. While it is clear that there is now a greater willingness than before to devote resources to safeguarding the environment, there are many other calls on the public and private purse (education, medical services, police, street lighting, and so on). At some point, a balance has to be struck between these competing demands.

Some approaches to pollution control (for example, best available technology) take no account of costs in relation to the environmental improvement that will result. This is wasteful of resources and cannot be sustained for long. A more sensitive, if more complicated, attack on pollution control is required.

Assessment of pollution hazards

It has become increasingly clear that there are no short cuts to safeguarding the marine environment. None of the more stringent measures proposed (and sometimes fiercely advocated) is sufficient in itself to protect the sea. In any case, we are now more aware of the interactions between land, sea, and air, and that it is ineffective to treat the problems in the sea in isolation. Environmental protection demands a holistic approach embracing all environments.

Environmental management requires a clear assessment of hazards. The hazard posed by any particular waste reaching the sea depends on its persistence, toxicity, and bioaccumulation potential. However, the actual impact it has once it is released into the environment depends also on its behaviour there: whether it remains in the water column or is sequestered in sediments, remains localized or becomes widely distributed. Finally, environmental protection must take account of variations in local circumstances, such as the capacity of the receiving waters to accept wastes without damage, and of critical environmental loads for particular substances.

Practical solutions to the problems created by waste disposal in the sea require a knowledge of what damage is caused by a discharge and the way in which the damage is caused. In addition to measuring the impact of a pollutant in the sea, we also have to assess its effects, that is, to decide upon its seriousness. A number of factors enter into this assessment, several of which cannot be measured in any direct way. The final assessment therefore involves judgements about which there may be legitimate disagreements, which is where much of the controversy about marine pollution arises.

THE SERIOUSNESS OF POLLUTION DAMAGE

Danger to public health

First and overriding priority is given to controlling a pollutant which poses a threat to human health. In the sea, conservative pollutants (most immediately mercury), pathogenic bacteria and viruses, and radioactivity are in this category. A distinction must be made between the **perceived risk** posed by these and the **actual risk**: the perceived risk reflects the gravity with which the hazard is viewed by the general public, and does not necessarily accord with the statistical reality, which is the actual risk. The perceived risk from radioactivity is very high, that from traffic accidents very low, although the actual risks are overwhelmingly opposite. For the most part, human exposure to harm from marine pollutants is through eating contaminated seafood, and humans generally appear to be particularly sensitive to tainting or stale flavours in seafood, or even to the suggestion that seafood is tainted, and reject it. The perceived risk to health from marine pollution is therefore very high and, for the most part, only very low actual health risks are acceptable.

This draws attention to the fact that, since the principal risk is from eating contaminated seafood, coastal communities whose diet includes a significant proportion of marine organisms are most at risk. This is particularly true of cumulative poisons such as mercury and it is not surprising that the best-known cases of mercury poisoning that originated from contamination of the sea should have involved fishing communities in Japan who subsisted almost entirely on food from the sea.

Mortality and morbidity

A good deal of pollution research is concerned with measuring the toxicity of substances which are (or may be) added to the sea. The nature and the limitations of toxicity tests are described in Chapter 4. They are valuable in giving warning of potentially damaging substances, they may help rank substances in order of toxicity, or, when damage is known to be caused by a complex effluent, they may indicate which of the constituents is most likely to be responsible for it. Generally, however, they have little value in assessing the impact of a pollutant in any particular situation.

Sub-lethal damage or delayed mortality, particularly if it is not revealed until the next or a later generation, escapes detection in conventional toxicity tests. While it may be concluded that greater mortality will be caused by a pollutant than that suggested by the LD_{50} measured in a short-term test, and the results of an investigation of sub-lethal effects provide a more urgent danger signal, they suffer the same shortcomings as other toxicity tests.

Whatever fears may be aroused by toxicity or sub-lethal studies in the laboratory, often the first sign, and sometimes the only visible evidence of pollution, is the sudden death of plants and animals in the affected environment. This may be only the tip of an iceberg; many more organisms may have been killed than were seen, particularly if there is an additional delayed mortality as a consequence of initial sub-lethal damage.

It is a relatively straightforward matter to assess the extent of damage when impounded fish and other species in mariculture are killed, because the fish or shellfish have a commercial value and compensation can be based, with reasonable justice, on the estimated cash value of the lost crop or the days of fishing lost. Very much greater difficulty is experienced when the casualties have no commercial value and attempts to quantify them in cash terms have, not unexpectedly, met with little success. A prime example was the assessment of damages when the tanker *Zoe Colocotroni* ran aground on the coast of Puerto Rico in March 1973, spilling part of its cargo of crude oil and contaminating some 8 hectares of mangrove and shoreline. The ship-owners were found guilty of negligence and in assessing damages the judge concluded that 92 109 720 animals had been killed and proceeded to assess their value by reference to the cost of replacements (or equivalent animals) if purchased from a biological supply house. This gave rise to the so-called 'ten-cents-a-barnacle' approach to assessing pollution damage. In the *Zoe Colocotroni* case, the court's judgement was later reversed by the appeal court on the grounds that the

destroyed plants and animals would have replaced themselves in the natural course of events within a relatively short time. Indeed, most had already done so by the time this decision was reached.

A far greater problem is created when the casualties can be regarded in some sense, as lost 'amenities'. Oiled sea birds and the death of seals and cetaceans fall in this category. While the strength of feeling—at least in some countries—caused by sea bird deaths from oil cannot be doubted, and is an indication of the strength with which this consequence of pollution is viewed, it has no scientific basis and cannot be quantified.

Population change

From a strictly biological point of view, when pollution causes the death of organisms, what matters is not the initial mortality, but the numbers and fate of the survivors.

Many marine organisms reproduce on a lavish scale: the American oyster produces 115 million eggs, and the cod 3–7 million eggs a year, breeding perhaps several times during its life. Their reproductive strategy is geared to enormous pre-adult mortality and for such animals pollution-induced mortality may be totally insignificant beside such natural mortality. Indeed, it has been calculated that a major oil spill at the time and place of the main fish spawning in the North Sea, using 'worst case' assumptions at each stage, would result in a small reduction in the overall yield for only one year, and that reduction would be difficult to detect against the background of the normal year-to-year variation in fish catches. Practical verification of this can be seen in the lack of any detectable impact of oil from the *Torrey Canyon* on the Cornish pilchard fishery despite the heavy loss of eggs and larvae (p. 51).

Species with a different reproductive strategy, that of producing few young which lead a protected early life, may not have the capacity to withstand heavy juvenile mortality. This is particularly true of species with a boreal reproductive pattern which is characterized by erratic breeding, low fecundity, slow growth and late maturation, but a long adult life. The breeding biology of sea birds such as guillemots exemplifies this arctic pattern (p. 44).

The consequences of heavy mortality on a population depend on the population dynamics of the affected species. Additional losses through pollution may cause a large and persistent population reduction or be totally obscured by natural mortality. While the immediate mortality may be damaging for commercial or other reasons in some circumstances, from a biological point of view, unless the mortality has an impact at the level of the population, it is of no consequence.

ASSESSMENT OF DAMAGE

Even when it is established that pollution-induced mortality has resulted in a population decrease in a species, assessment of the seriousness of this damage depends on what species is affected. A reduction of the population of a species of commercial importance or amenity value would be regarded much more seriously than a reduction in the population of an obscure worm. Apart from species that are of particular interest to humans, others which play a critical role in their ecological community, and whose loss has repercussions in the community, are of particular importance. The nature of the repercussions and the length of time they continue must also be assessed in making a final judgement of the seriousness of a pollution-induced change.

Key species in ecosystems

The reduction in the number of breeding auks in the colonies of southwest England and Wales during this century appears to have had no wider impact; and if the 1967 year-class of Cornish pilchards had been catastrophically reduced by *Torrey Canyon* oil and dispersants (see p. 51), although the commercial fishery might have suffered, there is no reason to suppose that it would have had widespread ecological repercussions. The loss of limpets (*Patella*) on some Cornish beaches, on the

other hand, led to major ecological change, with algal domination of the rocks for several years following the loss of the dominant herbivores (see p. 39). A similar ecological succession was observed in the sub-tidal rocks of a small bay in Baja California following the wreck of the *Tampico Maru* (see p. 40).

A potentially far more serious ecological change following the severe reduction in numbers of a key species has been recorded on the coast of Nova Scotia. Here there was a stable kelp bed ecosystem with *Laminaria* spp., as the main producers, responsible for about 90 per cent of the coastal production (most of which was exported offshore as detritus). The coastal area supported a population of herbivores, mainly the sea urchin, *Strongylocentrotus droebachiensis*, which was preyed upon and controlled chiefly by lobsters, *Homarus americanus*, and crabs, *Cancer*. The lobster population declined greatly after 1968 due, it was thought, to overfishing rather than pollution. With the reduction in predation, *Strongylocentrotus* increased greatly in numbers, and considerable areas of *Laminaria* forest were destroyed. It might have been expected that, having eaten out their food source, the sea urchins would then decline in numbers. They did not; instead, growth and gonad production were both reduced (Fig. 10.1) and the population remained large. Since areas denuded of *Laminaria* do not provide a suitable habitat for young lobsters and crabs, the population of these predators was not restored and the continued settlement of *Strongylocentrotus* remained unchecked. This urchin-dominated barren ground appears to be a new, stable configuration which could persist for a considerable time. The loss of primary and secondary production on substantial parts of the Nova Scotia coast which this represents is a serious threat to inshore, and possibly also to offshore, fisheries.

Recovery and duration of impact

Recovery of a damaged community begins as soon as the pollution is abated. Restoration is

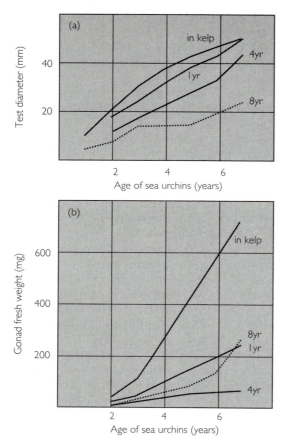

Fig. 10.1 Body growth and gonad development of sea urchins living in healthy kelp beds, and at sites where kelp has been destroyed by over-grazing one, four, and eight years previously.

from the vegetative growth of survivors *in situ*, immigration of mobile species from outside the area, and recolonization of the area by the distributive stages of the relatively sedentary species. Unless the damage has been only slight, for nearly all organisms (except plankton and fish) the last of these restoration sources is overwhelmingly the most important.

As the physical and chemical environment returns to normal, it will first become available to hardy species, and later to those which can tolerate only a degree of contamination. Species that breed, or otherwise have distributive stages throughout much of the year, are likely to recolonize before species with a

restricted breeding season, and re-establishment of the latter may be delayed for a year or more. The earliest colonizers which reproduce at most times of the year and are tolerant of adverse conditions are **opportunistic species** and they flourish in the absence of competition or predation. The polychaete *Capitella capitata* is such an opportunistic species in muddy substrata and because of its tolerance of organic enrichment is regarded as a pollution **indicator species**. As the environmental conditions improve, a succession of other species may follow until a stable, balanced community is built up.

Recovery of rocky substrata is more complex and less predictable, but follows a similar pattern of early colonization by opportunistic species and then the establishment of greater species diversity in the community. However, since rocky substrata are subject to greater wave action and water movement than soft substrata, the pollutant is more quickly removed from them and faunistic and floristic successions that can be observed during the recovery period are more concerned with the availability of colonizing stages and interactions between plants and animals, than with the steady improvement in environmental conditions.

With few exceptions, the most important consideration in assessing the environmental impact of pollution is neither the immediate mortality, nor the perturbation that this may cause to the population of the affected species or the community of which it is part, but the length of time the damage lasts. Damage that is restored within a few months is viewed quite differently from that which persists for years.

Experience following the wreck of the *Torrey Canyon* off the Cornish coast suggests that a considerable degree of recovery is possible on impacted rocky shores and is achieved within two or three years, but where dominant herbivores were killed and *Fucus* became established, this inhibited re-establishment of limpets for several years. Restoration of the former community took seven or eight years, and for some years after that subtle impoverishment of the fauna could still be detected. Recovery of damaged benthic communities is slow, possibly because, in those physically quiet conditions, the pollutant is slow to be removed. An ecological succession continued for six years before the sub-tidal benthic community began to resemble that of unaffected sites following the reduction in oily discharges from a Finnish refinery. A similar period elapsed following an oil spill in Buzzard's Bay, Massachusetts, and even then the benthic fauna showed fluctuations and instability. Recovery of a benthic community in a sea loch in west Scotland, after cessation of discharges of organic waste from a pulp mill, followed a similar course and time-scale.

In temperate waters, severe pollution damage may be restored within two to three years, but in most situations it is more likely that six to ten years will be required before something that may be regarded as a 'normal' community is re-established. Some investigators claim that subtle damage persists for even longer periods. We know much less about arctic and tropical waters, but there is reason to expect that recovery processes will be different in those waters from processes observed in temperate regions.

Arctic fauna and flora are characterized by erratic reproduction and very low fecundity. The great majority of marine invertebrates lack distributive phases and either brood eggs and young, or lay eggs in a protective attached capsule. In these circumstances, if a substantial area is damaged by pollution, re-colonization will be very slow indeed because of the relative immobility of most marine organisms and their young. Terrestrial arctic habits are known to be extremely fragile and slow to recover, and the marine environment is likely to be the same. The time required for substantial restoration of pollution damage is unknown; 100 years has been suggested and could well be correct.

In the tropics, it is common for marine organisms to have pelagic larvae or other distributive phases, and breeding seasons are longer. Many species which breed only annually in temperate waters breed several times a year in the tropics, or tropical species have a much shorter generation time than their temperate water counterparts. It is also common for tropical species to have briefer lives than their temperate water or polar relatives. There is a much faster turnover in tropical ecosystems and colonizing forms are available at most, if not all, times of the year. Pollution damage is therefore rapidly made good and, instead of the decade that might be required for complete recovery in temperate waters, one to two years is probably a more realistic period in the tropics. There is some concern that mangrove forests and coral reefs may be more vulnerable tropical marine habitats than most, and not recover so readily from damage. Both mangroves and reef-building corals provide the physical conditions on which the large and varied associated fauna depend. Destruction of these structural elements would therefore have far-reaching consequences which might be prolonged. Whether these fears are justified remains to be discovered as more experience is gained of pollution impact in tropical waters—as unfortunately seems likely to happen.

Evaluation of change

Materials accidentally or deliberately discharged into, or otherwise reaching the sea, may cause greater or lesser biological change, of long or short duration. In evaluating this pollution impact, it is necessary to decide whether the change is desirable, undesirable, or neutral. There is a temptation to regard any change as necessarily undesirable, as most of the examples cited in this book would probably be judged. However, if the ways in which the change is undesirable have to be enumerated and evaluated, it becomes obvious that not all change is necessarily for the worse.

1. The reduction or disappearance of auk colonies in Brittany and southwest England is biologically neutral: they are fringe populations which appear to be contracting as part of a northward recession of these birds in

response to climatic change. This is small consolation to bird watchers in the affected areas, for whom the decline in local sea bird populations is a loss of amenity. It has been possible to restock a colony of puffins on the coast of Maine with Canadian birds, and a similar restocking programme with Faeroese birds has been attempted in the Sept Iles Nature Reserve on the coast of Brittany. Positive conservation measures of this kind may, with difficulty, maintain a population of a formerly abundant species in protected areas like nature reserves, but cannot reverse a climatically induced trend.

2. The change, lasting several years, from barnacle domination on some rocky beaches on the Cornish coast to an algal covering, as a consequence of killing grazing limpets with toxic oil dispersants, constituted the greater part of the ecological 'disaster' following the wreck of the *Torrey Canyon*. The affected shores were in an area heavily dependent upon tourism and it is true that most tourists prefer barnacle-covered rocks to a beach covered with a dense growth of wet, slippery seaweed. From that point of view, the change was undesirable. It is less clear in what biological sense an alga-dominated beach, with its associated fauna, is more or less desirable than a *Patella–Balanus* assemblage.

3. The prolonged discharge of china clay waste into St Austell Bay in south Cornwall (p. 117) converted a considerable area of seabed from rock and gravel to a soft substratum. A hard-bottom fauna was replaced by a soft-bottom fauna of polychaetes and bivalves, with flatfish preying on them. This was damaging commercially because of the greater market value of lobsters than flatfish. From a biological point of view, the change in the nature of the ecosystem was probably neutral.

4. The replacement of one bivalve, *Nucula*, by another, *Abra*, on the sewage sludge dumping grounds in the German Bight (p. 21) may similarly be regarded as biologically neutral and, indeed, appears to have had no harmful economic consequences. Periods of anoxic bottom waters in the Bight were both bio-logically harmful and damaging to fisheries, but it is not certain that the input of sewage sludge contributed significantly to this.

5. Sea urchin domination of the Nova Scotia coast and destruction of the *Laminaria* forest as a consequence of over-fishing lobsters has resulted in a tremendous loss of production and this appears to be a long-term, if not permanent, change. Failure of the lobster fishery to recover after lobster fishing stopped is commercially harmful. The loss of primary production is biologically damaging and, if it leads to an impoverishment of offshore waters, may inflict further commercial damage on offshore fisheries.

6. Chronic discharge of refinery effluent at Fawley, in Southampton Water, caused severe environmental deterioration measured by a reduction in species diversity in the salt marsh over which the oily effluent flowed (p. 44). Biologically, this change was very damaging, but in this case no human interest appears to have suffered as a result of it.

It would be possible to multiply these examples, but it is evident that environmental change resulting from the addition of wastes to the sea requires dispassionate consideration before the extent and nature of the damage it represents can be assessed. As shown below, ecosystem changes on at least the same scale, and often of a similar nature, to those induced by pollution, sometimes occur from natural causes. These help to put pollution damage in perspective and have to be taken into account when making a final assessment of the significance of an observed ecosystem adjustment.

PROBLEMS OF MEASURING POLLUTION IMPACT

Environmental science and many matters relating to pollution in the sea are contentious, and reputable scientists may arrive at quite different evaluations of marine pollution. When scientists disagree in this way it is usually because the question posed is not clearly defined, or because the data on which the answer is based are inadequate.

Much of the effort in pollution research has been directed towards collecting the following data:

(1) measuring the concentration of contaminants in the water column, sediments, and marine organisms;

(2) studying the behaviour, degradation, or modifications of these contaminants once they reach the sea;

(3) discovering their transmission through food webs;

(4) measuring their effects through toxicity tests of varying degrees of refinement and sophistication, and by studying the physiological, developmental, or behavioural responses of organisms exposed to them.

No-one would deny the importance of knowing the quantity and source of inputs to the marine environment and their behaviour after they get there, but the fact remains that much of the enormous volume of quantitative data that is being accumulated cannot be applied to any real-life situation.

When is recovery complete?

One area of disagreement among pollution scientists is the length of time required for the recovery of a damaged ecosystem; this is because of uncertainty about when recovery is complete.

Like the auks in southern England, a variety of British marine organisms are near the fringe of their geographical range. If a population of such a species is destroyed by pollution, its replacement is a matter of chance and, as in the case of the sea bird colonies, it may not be replaced at all. The British Isles, lying near the southern limit of arctic and subarctic species and the northern limit of warm-water species, has many such precarious fringe populations. Ecologically, such populations are of little significance, though they may have considerable scientific interest, and it would be unrealistic to wait for the reappearance of such a species in a community before regarding recovery complete.

It is also unnecessary to wait for a balanced age structure to be re-established in a damaged community. Lobsters may live for 10 or 20 years; it would obviously be unrealistic to wait 20 years to find a few old specimens in the population. In any case, many marine species do not show regular recruitment and an actuarial age struture. Populations are often dominated by a particular year-class and may show little new recruitment for several years (Fig. 10.2).

In other words, recovery is marked by the re-establishment of a healthy community, but not necessarily one that is identical in composition and age structure to the community that was present before the pollution damage. Unfortunately, we cannot define a healthy community in any rigorous way and hence there is room for considerable disagreement about the recovery of an ecosystem.

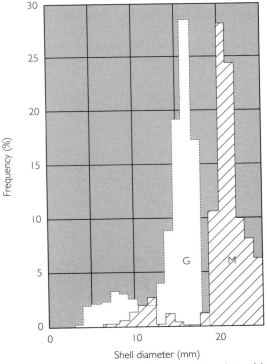

Fig. 10.2 Size distribution of the intertidal gastropods *Gibbula* (G) and *Monodonta* (M) in a population on the Pembrokeshire coast, October 1981.

Natural population fluctuations

It is now becoming clear that there are very wide population fluctuations in marine ecosystems from purely natural causes, although the latter are rarely understood in detail. Not only is it extremely difficult to distinguish pollution-induced change against this fluctuating background, but sometimes the natural fluctuations are sufficient to produce a major ecological change comparable to that resulting from what would be regarded as a catastrophic pollution incident. The same problems are encountered in 'baseline' studies, many of which are conducted for too short a period to reveal the full range of fluctuations in the ecosystem, and therefore have little value.

The extent of this problem was dramatically shown by Dr J. R. Lewis following several years' continuous study of a rocky coastal area in northeast England. This beach, with moderate wave exposure, is equally suited for colonization by *Mytilus*, *Fucus*, and *Patella*, with barnacles, each excluding the others. Its population changes are outlined below:

1965–6: A cold winter stopped the growth of the 1965 class of the carnivorous gastropod *Nucella lapillus* (the dog whelk), which was eaten in large numbers by the purple sandpiper, *Calidris maritima*.

1966–7: The winter was milder and the 1966 class of *Nucella* fed and grew continuously, becoming too large to be eaten by the sandpipers. A large population of *Nucella* survived the winter, feeding on *Mytilus* and seriously reducing its numbers, so making space available for colonization by *Balanus*.

1967: Before *Balanus* had a chance to settle, there was an exceptionally heavy spatfall of *Mytilus* which smothered all other species on the rocks, filled the gaps, and excluded *Balanus*. The *Mytilus* were so dense that there was severe competition for space, resulting in the mussels growing in unstable hummocks.

1969: In March, storms swept many *Mytilus* away, so leaving space once more for a *Balanus* settlement in June. But, because the dense growth of *Mytilus* had excluded the limpet *Patella*, diatoms and ephemeral algae developed quickly and abundantly and prevented the settlement of *Balanus*. The rocks then became dominated by algae.

The important role of *Patella* in determining the rocky shore community is revealed by prolonged studies of the abundance of *Balanus* or *Mytilus* at individual sites in the same area (Fig. 10.3). The wide and erratic fluctuations are due to the operation of a variety of influences, but an important factor is the success of *Patella* recruitment. The success of recruitment in any year is not much affected by the number of gametes produced, but is probably influenced by the climatic factors during the first few weeks after the young have settled on the shore (Fig. 10.4).

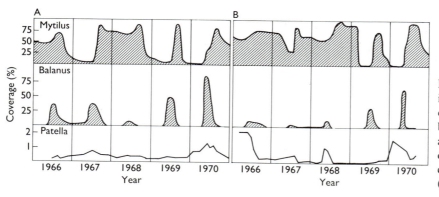

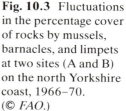

Fig. 10.3 Fluctuations in the percentage cover of rocks by mussels, barnacles, and limpets at two sites (A and B) on the north Yorkshire coast, 1966–70. (© *FAO*.)

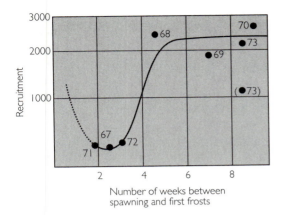

Fig. 10.4 Relation between the recruitment of limpets, *Patella vulgata*, and the duration of the frost-free period after spawning for the years 1967–73. (*Cambridge University Press.*)

Faunistic changes, also apparently due to the influence of climatic factors, have been detected in the benthic soft-bottom fauna off the Northumberland coast (Fig. 10.5). The assemblage of polychaetes in the community is a balanced one, but the relative abundance of different species changed as the average winter sea temperature increased from 6.25 °C in 1965–70 to 6.75 °C in 1971–6. Some species increased in abundance, others decreased, and a few were unaffected. No species was eliminated and there seems to have been little change in community production; the available energy was simply transferred from one suite of species to another which thrived better in response to the environmental change. Presumably if the former environmental conditions were restored there would be no obstacle to a return

Gaining species 1972-6 (G)		Losing species 1972-6 (L)		Neutral species 1972-6 (Nt)
Species	Gain m⁻²	Species	Loss m⁻²	Species
Paraonis gracilis	238	Spiophanes bombyx	219	Ampharete finmarchica
Prionospio malmgreni	144	Chaetozone setosa	129	Glycera rouxi
Myriochele oculata	129	Ampelisca tenuicornis	112	
Thyasira flexuosa	115	Amphiura filiformis	50	Goniada maculata
Owenia fusiformis	104	Nephtys hombergi	31	
Magelona minuta	100	Scoloplos armiger	26	
Tharyx multibranchiis	27	Rhodine gracilior	24	
		Ammotrypane aulogaster	17	

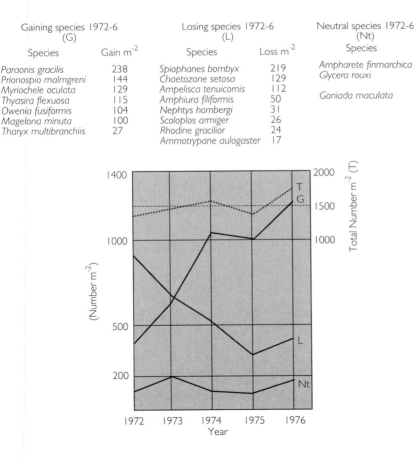

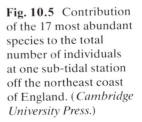

Fig. 10.5 Contribution of the 17 most abundant species to the total number of individuals at one sub-tidal station off the northeast coast of England. (*Cambridge University Press.*)

to the original balance of species in the community.

Long-term trends in the abundance of various species of north Atlantic zooplankton were detected in a 22-year Continuous Plankton Recorder Survey from 1948 to 1969 (Fig. 10.6). Some species, such as *Pleuromamma borealis* (a) and *Euchaeta norvegica* (b) showed an upward trend; *Spiratella retroversa* (j) and two calanoid (k) species showed a significant downward trend; *Temora longicornis* (d) and *Clione limacina* (e) showed no trend at all. All the species were subject to erratic and sometimes very large fluctuations in abundance—some fluctuations extending over several years—and the long-term trends would not have been detected without this very long time-series of observations. It has not been possible to relate these long-term fluctuations and trends to any known climatic or human factors.

Historical records show that fish stocks have undergone the most dramatic changes. Good and bad periods for a number of

herring stocks can be traced back for 600 years (Fig. 10.7) and seem to be related to climatic changes which were reflected also in the extent of the Arctic ice-cap, severity of winters, and similar phenomena. Shorter-term, wide fluctuations in the abundance of cod in the northeast Atlantic are known to have occurred during the last century and have been reflected in landings in the Lofoten Islands between 1860 and 1955, and in total landings for the northeast Atlantic cod fishery between 1945 and 1978 (Fig. 10.8). The weight of fish landed depends on the amount of effort put into catching them, as well as upon their abundance, and there is no guarantee that fishing effort remained constant throughout these periods. The observed changes in landings have been so abrupt and so great, however, that enormous changes in fishery practices would have had to take place to account for them. Whatever changes of fishing effort may have taken place, they were certainly not on a great enough scale to cause this and there can be no doubt that the size of

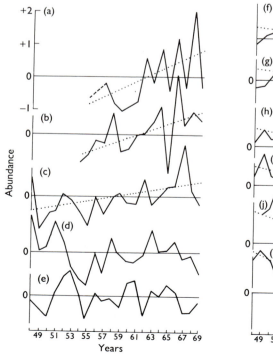

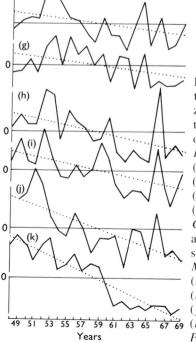

Fig. 10.6 Changes in the abundance of zooplankton species in the North Atlantic. The organisms are: (*a*) *Pleuromamma borealis*; (*b*) *Euchaeta norvegica*; (*c*) *Acartia clausi*; (*d*) *Temora longicornis*; (*e*) *Clione limacina*; (f) *Calanus helgolandicus*; and *C. finmarchicus* stages V and VI; (*g*) *Metridia lucens*; (*h*) *Candacia armata*; (*i*) *Centropages typicus*; (*j*) *Spiratella retroversa*; (*k*) *Pseudocalanus* and *Paracalanus*. (© *FAO*.)

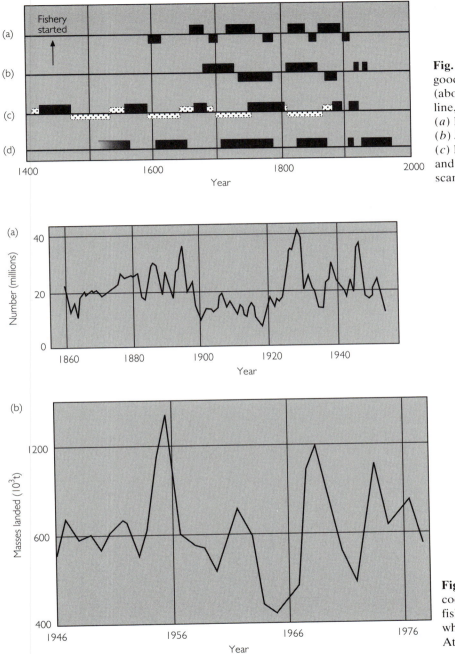

Fig. 10.7 Periods of good and bad years (above and below the line, respectively) for (*a*) Hokkaido herring, (*b*) Japanese sardine, (*c*) Bohuslan herring, and (*d*) Atlantoscandian herring.

Fig. 10.8 Landing of cod (*a*) in the Lofoten fishery and (*b*) in the whole northeast Atlantic fishery.

the annual catches gives a fair guide to the abundance of fish.

Several lessons can be learned from these studies of natural ecosystem fluctuations. Had a fisheries biologist begun to monitor fish stocks at the beginning of a downturn, he could well have seen, over a period of five years or more, an apparently catastrophic and irretrievable collapse of the fishery. In the present climate of opinion, the collapse would no doubt be attributed to pollution, over-fishing, or some other human activity. A longer-term perspective indicates differently.

A survey made at a place which receives an effluent and is suspected of being polluted by it may well show signs of some change in the ecosystem. It would be natural to conclude that the local contamination caused the change but, unless the survey has been carried out over a long enough time and on a suffi-cient geographical scale, it is impossible to be sure. The change that has been detected may have been caused by pollution, but equally it may be part of a wider, natural change.

Even the worst pollution damage is no greater than, and often not much different from, ecosystem changes caused by natural events. A pollution-induced change is avoid-able whereas a natural one is not, but fears that pollution is causing irreversible damage or that life in the seas is being destroyed appear to be exaggerated.

Genetic variability

Marine invertebrates, in general, show a high degree of genetic variability and fish are genetically the most varied of all vertebrates. Views about the adaptive significance of this condition are still largely speculative, but there is some evidence that, for some organ-isms, genetic variability is an adaptation to surviving in unpredictably variable environ-ments.

Estuaries undergo regular, predictable, short-term fluctuations with the tidal regime, and the organisms living there have been able to evolve physiological tolerance or regu-latory mechanisms to survive the changes in their physical and chemical surroundings. This is not possible in environments where the changes are unpredictable or occur at intervals comparable to the life-span of the individual. If the species is genetically varied, however, there may be on-the-spot selection of whichever genotype is adapted to the conditions prevailing at the time. Some individuals do not survive a change in environmental conditions but are replaced by a different genotype that can do so.

Many marine habitats, instead of showing temporal variability where conditions change from time to time, show spatial variability and are a mosaic of microhabitats. In each micro-habitat, there may be selection of larvae after they have settled, and the survivors may be those genotypes that are adapted to a some-what different set of conditions at each site.

There is great variation in size and struc-ture of the stipe of *Laminaria* and while to some extent this is the result of a phenotypic response to water movements, there is also a strong genetic selection of different kinds of plants with different growth characteristics at different sites representing different micro-habitats.

There are genetic differences affecting the withdrawal reflex of the small tubicolous polychaete, *Spirorbis borealis*, in open coast and sheltered rock pools only a few hundred metres apart, and in sunlit and shadowed pools.

Littorina populations show genetic differ-ences with respect to shell shape, to withstand wave action and desiccation.

Oysters, *Crassostrea virginica*, from differ-ent sites show genetic differences in growth rate and salinity tolerance.

Selection of successful genotypes may also be one method of adjusting to contamination of the environment. There is, for instance, very rapid selection of alleles producing resistance to water-soluble components of crude oil in the copepod *Tisbe* (Fig. 10.9). Organisms living in Cornish estuaries con-taminated with copper accumulate very large quantities of the metal (Fig. 5.6, p. 77), that would be lethal to members of the same species living in uncontaminated habitats. It is likely that copper-tolerant genotypes have been selected.

The same strategy for survival may underlie the success of opportunistic species, which show rapid population changes and are able to respond to environmental disturbances. The polychaete, *Capitella capitata*, for example, is now known to be a complex of sibling species, several of which are them-selves genetically very variable. With a repro-

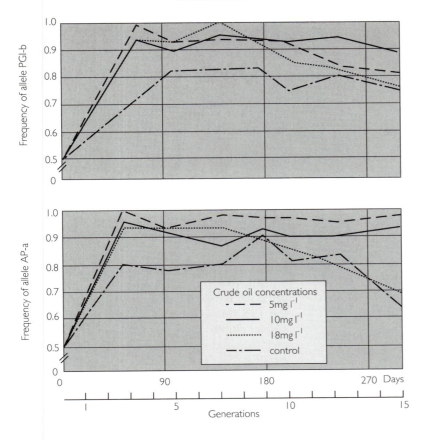

Fig. 10.9 Changes of gene frequency for two alleles in the copepod *Tisbe holothuriae* in oil-contaminated water.

ductive pattern which ensures that colonizing young are available at most times of the year, one or other of the siblings or genotypes will be among the first animals to re-occupy a polluted area, and they will then flourish in the absence of competition.

One strategy for survival in the marine environment thus appears to be not so much an efficient machinery for tolerance or regulation, so that the individual withstands environmental change, but a genetic flexibility that ensures that there is a range of genotypes available, so that one can be selected to flourish in whatever environmental conditions are encountered. The existence of clusters of sibling species in *Capitella*, *Ophryotrocha*, and other polychaetes (and no doubt more are yet to be discovered) is an extension of the same phenomenon.

This phenomenon may explain the surprising robustness of many marine eco-

systems to substantial levels of contamination arising from human activities. There may well be subtle adjustments, but if they involve the replacement of one genotype or sibling species by another, they are unlikely to be detected by ecologists. This genetic flexibility poses practical problems for environmental studies and even greater difficulties for laboratory studies of pollution effects.

The selection of genotypes that will thrive in local conditions which vary within the space of a few metres, or even on the opposite sides of the same boulder, results in a mosaic. Figure 10.10 shows the growth rates of the algae *Laminaria* and *Agarum* at three places on the Nova Scotia coast of Canada. Although several hundred individual plants were measured in each sample, there are still standard deviations of ±100 per cent of the mean values. With such inherent variability in the system, very large numbers of samples

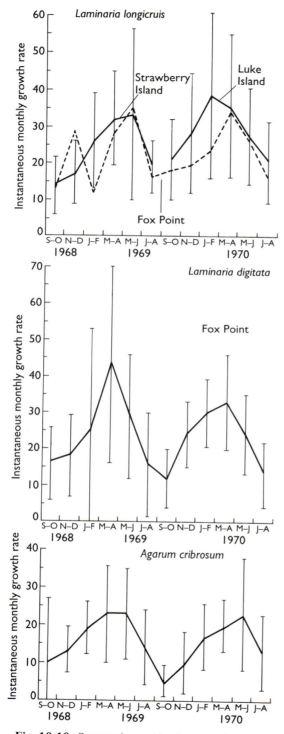

Fig. 10.10 Seasonal growth of seaweeds (means and standard deviations) on the Nova Scotia coast.

and measurements are needed to get valid results from an investigation. This is very time-consuming and has rarely been attempted. The variability is usually ignored or concealed, and because of this it is possible that many published results are meaningless.

The extreme complexity and variability of the natural environment makes pollution studies and monitoring in the field very difficult. This has led to a greater concentration on laboratory studies, but there the implications of the natural variability of marine organisms are even more serious. The whole philosophy of experimental biology is to obtain reproducible results under known, controlled conditions. Experimentalists expect to get low standard deviations and, indeed, experimental results with a standard error of ±100 per cent of the mean (as with the *Laminaria* growth rates) would be quite unacceptable. Genetic variation between the experimental organisms is to be avoided if at all possible, and the use of the laboratory white rat in genetically pure strains is an answer to that problem. In experimental pollution research, the equivalent has been the use of:

(1) a small number of 'standard' animals, as in, for example, toxicity tests; laboratory-cultured animals in preference to natural populations;

(2) genetically pure strains of *Ophryotrocha* developed in Sweden;

(3) cloned hydroids *Campanularia* which are, of course, genetically identical and have been used for the bioassay of pollutants in low concentrations.

This approach is in the best traditions of experimental biology, but if, as it now appears, genetic variability is an essential feature of natural populations and on-the-spot selection of successful genotypes the key to their survival in varying environments, experiments on genetically homogeneous animals may completely falsify the natural situation. It makes it virtually impossible to translate the results of such laboratory studies into real-life terms. Due to our inadequate knowledge of the response of marine populations and communities to toxins and other stresses in the natural environment, experimental results at best cannot be interpreted and, at worst, may be totally misleading.

THE PLACE OF SCIENCE IN POLLUTION ASSESSMENT

In all science, the right answers emerge only if the right questions are asked, and the answer to one question should lead to further questions. Pollution science is a little more complicated than that. Effluents discharged to the sea, or reaching it by a circuitous route, constitute pollution only when they have some undesirable effect, but that raises many questions about what is undesirable or acceptable, and to whom. Those questions are political, not scientific. The primary scientific questions relate to environmental change. Before a sensible political decision can be taken, it would be desirable to discover the nature, extent, and duration of an environmental change, and to identify the causative agents and possible remedies. That takes time, and political decisions often have to be taken without waiting for the scientific evidence to be gathered. Scientists must therefore be prepared to give a balanced appraisal on the basis of what is already known. Even when the necessary information is available, there is no guarantee that wise political decisions will be taken, but if scientists cannot even begin to provide it, the chance of adequate and economical remedial action being taken is remote indeed.

Pollution control measures are certainly not always undertaken on a rational basis. Because of the uncertainties of the situation, there is a temptation to play safe, and remedial action is sometimes instituted before it is known whether a waste product is damaging in any material way, or if the action taken will have a beneficial effect. Aroused, but ill-formed, public opinion may insist that something be done to remedy a situation and

an industry or government body be obliged to undertake completely useless activities so that they can be seen to be active. Such activities may be harmless, but they are expensive and wasteful of resources that might be better spent for some other purpose. In other respects, irrational attempts at pollution control may be positively harmful. Awareness of possible dangers stemming from pollution of the sea has encouraged the view that nothing should be discharged or dumped into it, but human society creates wastes and these have to go somewhere. Care must be taken to ensure that in solving one problem we are not simply creating a worse one in a different environment. There is obviously a great need for scientific judgement in assessing pollution dangers and prescribing remedies, even if the scientist is only one contributor to the debate.

The pollution scientist is still faced with a number of fundamental problems. Some, possibly all, marine ecosystems show wide, erratic, natural fluctuations caused by climatic or biological events which we are only just beginning to understand. They can be on a scale comparable to the effects of the most severe pollution damage, sometimes, as in the case of the loss of dominant herbivores, producing exactly the same kind of change or, as in the case of the exceptionally severe and prolonged winter of 1962–3 in northwest Europe, causing even greater and more widespread change. These are only the most conspicuous natural fluctuations and no doubt there are constant adjustments in ecosystems on a much smaller scale. Against such a background it is difficult to detect pollution-induced changes unless they are very obvious and large, or are clearly related to a single-point input. To get early warning of pollution damage, particularly from diffuse inputs, requires a much more detailed and profound knowledge of ecosystem structure and performance than we usually have. In addition to this need, it must be remembered that the collection of data is complicated by the genetic variability of organisms which is now coming to light. To collect reliable data in the field, much larger numbers of samples and measurements are needed than would normally be contemplated. A concentration on the simpler and more controlled situation in the laboratory is no solution to this problem if, as sometimes seems likely, the laboratory studies lack realism and give misleading results or, at best, yield results which can be applied to a real-life situation only with great difficulty.

Because of these difficulties it is understandable that scientists have turned attention much more to short-term surveys, and to experimental studies and chemical monitoring of the pathways, concentrations, and fates of additions to the marine environment for which sensitive analytical equipment can be used. Unfortunately, it is now becoming clear that, although we need the results of these kinds of investigations, we also need a much more sophisticated approach to the problems of waste disposal. We need to know the safe limit of effluent discharges to particular areas of the sea, the response of marine ecosystems to stress, and the environmental cost of alternative methods of waste disposal. We live in a finite world and we can no longer consider the protection of the marine environment in isolation from pressures on other environments.

FURTHER READING

There is an enormous scientific literature, of variable quality, on matters relating to pollution in the sea, and a wide range of popular books on environmental matters, although the latter are mostly concerned with putting forward a particular point of view. The following titles give more technical detail on subjects discussed in this book and guide the reader to important works in the scientific literature.

Abel, P. D. and Axiak, V. (ed.) (1991). *Ecotoxicology and the marine environment* (Ellis Horwood). In this book, experts describe recent developments in the measurement of the toxicity of pollutants to marine organisms, and also the application of toxicological data to marine pollution control.

Clark, R. B. (ed.) (1987). *The waters around the British Isles: their conflicting uses.* (Oxford University Press). This book examines in detail commercial activities in and around western European waters, conflicts of interest that arise, and the national and international legal regimes that regulate these activities.

Cormack, D. (1983). *Response to oil and chemical pollution.* (Applied Science, London). At the time this book was written, Dr Cormack was Chief Scientific Adviser to the UK Marine Pollution Control Unit, responsible for co-ordinating the response to major pollution accidents. The book deals comprehensively with the merits of the various methods employed to deal with oil and chemical spills.

Furness, R. W. and Rainbow, P. S. (ed.) (1990). *Heavy metals in the marine environment* (CRC Press, Boca Raton, Florida). An examination of current information about metal pollution in the sea.

Geraci, J. R. and St Aubin, D. J. (ed.) (1990). *Sea mammals and oil: confronting the risks* (Academic Press, London). Ten experts examine the effects of spilled petroleum on seals, whales and dolphins, sea otters, polar bears, and manatees.

Gerlach, S. A. (1981). *Marine pollution: diagnosis and therapy* (Springer, Heidelberg). This is an English translation of a book first published in 1976, with some updating. It is useful for giving the German experience of marine pollution not usually found in English books, but there have been developments in German attitudes and improvements in knowledge since the book was published and it is now a little dated.

GESAMP (1991). *The state of the marine environment* (Blackwell Scientific, Oxford). GESAMP is the Joint Group of Experts on the Scientific Aspects of Marine Pollution set up by the United Nations. Here international experts report on the present state of the world's oceans and the threats posed by different kinds of wastes.

Greenpeace (1987). *Coastline: Britain's threatened heritage* (Kingfisher Books, London). A beautifully illustrated and very readable tour around the British coastline, examining local environmental problems in a balanced way.

Johnson, R. (ed.) (1976). *Marine Pollution* (Academic Press, London). This is solid reading, but detailed and, despite its age, still authoritative. The chapters on fisheries, oil pollution, metals, and sea birds are particularly good.

Langford, T. E. L. (1990). *Ecological effects of thermal discharges* (Elsevier, London). This book reviews the biological impact of hot water discharges to the sea, particularly from thermal power stations. It includes an extensive bibliography.

Marine Pollution Bulletin (Pergamon, Oxford).

This monthly journal contains news and articles about marine pollution around the world. Many of the articles are designed for a non-technical audience.

Newman, P. J. and Agg, A. R. (ed.) (1988). *Environmental protection of the North Sea* (Heinemann, London). This volume includes contributions and discussions at an international conference on the North Sea in the run-up to the Ministerial meeting held in November 1987. The coverage of subjects and attitudes is comprehensive and authoritative.

Pentreath, R. J. (1980). *Nuclear power, man and the environment* (Taylor and Francis, London). A useful introduction to radiobiology and radiological protection, although only a small part of it is concerned with the marine environment.

Royal Commission on Environmental Pollution. The reports of the Royal Commission give clear, non-technical accounts of the environmental problems under review, as well as discussing remedies. All the reports are very thorough and authoritative. The following deal, at least in part, with the marine environment. Third report (1972) *Pollution in some British estuaries and coastal waters.* Sixth report (1976) *Nuclear power and the environment.* Eighth report (1981) *Oil pollution of the sea.* Tenth report (1984) *Tackling pollution: experience and prospects.* Twelfth report (1987) *The best practicable environmental option.* (All published by HMSO, London.)

Spellerberg, I. F. (1991). *Monitoring ecological change* (Cambridge University Press). This book discusses various monitoring techniques and assesses currently used monitoring programmes.

INDEX